T0335765

Agricultural Implications of the Fukushima Nuclear Accident

Tomoko M. Nakanishi • Keitaro Tanoi
Editors

Agricultural Implications of the Fukushima Nuclear Accident

The First Three Years

Editors
Tomoko M. Nakanishi
Graduate School of Agricultural and Life Sciences
The University of Tokyo
Bunkyo-ku, Tokyo
Japan

Keitaro Tanoi
Graduate School of Agricultural and Life Sciences
The University of Tokyo
Bunkyo-ku, Tokyo
Japan

ISBN 978-4-431-55826-2 ISBN 978-4-431-55828-6 (eBook)
DOI 10.1007/978-4-431-55828-6

Library of Congress Control Number: 2015960944

Springer Tokyo Heidelberg New York Dordrecht London

Printed on acid-free paper

Springer Japan KK is part of Springer Science+Business Media (www.springer.com)

Foreword

Four years have passed since the Tohoku earthquake struck on March 11, 2011. With a focus on Fukushima Prefecture, the recovery of regions that incurred damage from the nuclear power plant accident is continuing, and citizens are returning to their homes. For the agriculture, forestry, and fishing industries in Fukushima Prefecture, the nation and the municipality are utilizing methods that assist exclusion of radioactive materials, e.g., potassium fertilization. Moreover, there are adequate countermeasures against radiation exposure in Fukushima Prefecture; for example, entire crops of rice are inspected prior to shipment. Through these countermeasures, all of the products that are introduced in the market are managed so that they are below the standard limits for radionuclides. However, it seems that it will take a considerable amount of time for decontamination, for the decay of radioactive material, and for consumer anxiety about agricultural, forestry, and fishing products to be dispelled.

Immediately after the accident, the Graduate School of Agricultural and Life Sciences at The University of Tokyo created an independent team of specialists in a wide variety of areas including soil, vegetation, animal life, fishing, and forestry. This team entered the affected areas right after the accident and proceeded with research and studies. In addition, it was important to the Graduate School of Agricultural and Life Sciences that the results of this research and these studies should be useful to the recovery of the affected area; therefore, we have worked to officially announce these results. For example, there have been 11 meetings to report research results since November 2011. The objective of these meetings was to provide a simple explanation of the results of the studies and research so that the general public could understand.

Right after the first book about the research in Fukushima was published by Springer, an easy-to-understand book was published in Japanese. That book was published to allow a wide range of ordinary people to correctly understand the impact of radioactive material on agricultural, forestry, and fishing products and the countermeasures taken against radiation exposure.

It appears that it will still take quite some time for the agricultural, forestry, and fishing industries in Fukushima Prefecture to recover, but the Graduate School of Agricultural and Life Sciences at The University of Tokyo will continue to support the recovery of these industries in the future.

Ken Furuya

Professor
Dean, Graduate School of Agricultural and Life Sciences
The University of Tokyo
Bunkyo-ku, Tokyo
Japan

Preface

More than 4 years have passed since the Fukushima Daiichi nuclear power plant accident. Even now, some 110,000 people who were evacuated from the area are in locations away from where they used to live. Right after the nuclear accident, 40–50 academic staff members of the Agricultural Department of The University of Tokyo began to study the movement of radioactive materials emitted from the nuclear reactor, because most of the contaminated area in Fukushima is related to agriculture. Researchers are still continuing their work to find out what effects the accident has had on agricultural fields. Our Graduate School of Agricultural and Life Sciences (The University of Tokyo) comprises many research areas and there are many facilities attached to the School, such as meadows, experimental forests, and farm fields. Together with these facilities, many on-site research studies have been conducted in Fukushima. The projects now ongoing can be illustrated as follows.

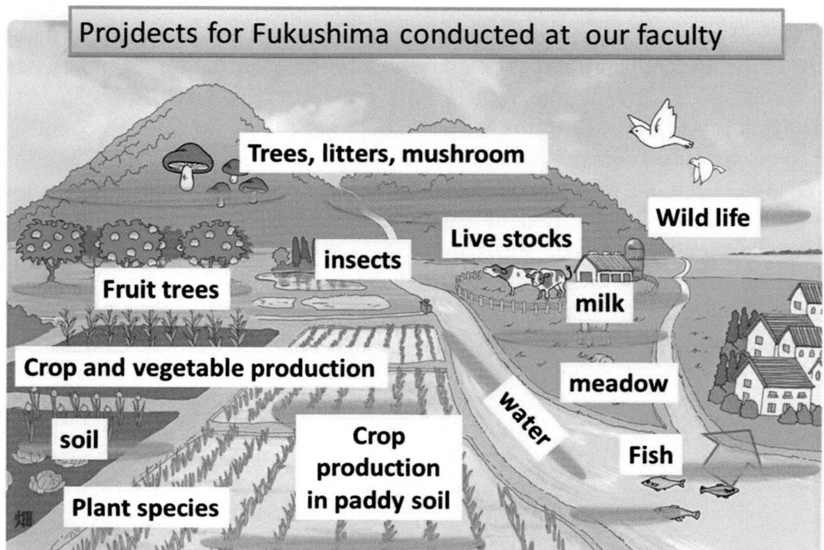

Through our activities many scientific findings have been collected. Based on these scientific data we started lecture classes for students as well as for the general public to explain the actual movement of radioactivity in nature, along with basic knowledge about radioactivity. So that students can experience radioecology in nature for themselves, we are periodically holding open lectures in the contaminated fields or mountains.

The first collection of papers on this subject was published by Springer Japan in 2013. It was made available as an open access book, free to download, so that the results of the research and studies could be widely shared with foreign and domestic researchers. We received an extremely large amount of feedback from that effort. There were many queries from researchers both inside and outside the country, as well as words of encouragement.

We are continuing with our studies, and we have decided to publish the present volume as a second collection of papers summarizing the subsequent research results. This collection will show more detailed results about the transfer of radioactive material into agricultural products and the movement of radioactive materials within environments like forests. We hope that this will be of use to everyone concerned in the same way that the first collection of papers was.

Tokyo, Japan Tomoko M. Nakanishi

Contents

1 **An Overview of Our Research** 1
Tomoko M. Nakanishi

2 **Monitoring Inspection for Radioactive Substances in Agricultural, Livestock, Forest and Fishery Products in Fukushima Prefecture** 11
Naoto Nihei

3 **Rice Inspections in Fukushima Prefecture** 23
Naoto Nihei

4 **Cesium Accumulation in Paddy Field Rice Grown in Fukushima from 2011 to 2013: Cultivars and Fertilization** 33
Yoshihiro Ohmori, Nobuhiro Tanaka, and Toru Fujiwara

5 **Physiological Verification of the Effect of Potassium Supply on the Reduction of Radiocesium Content in Rice Grain** 45
Natsuko I. Kobayashi

6 **Consecutive Field Trials of Rice Cultivation in Partially Decontaminated Paddy Fields to Reduce Radiocesium Absorption in the Iitate Village in Fukushima Prefecture** 55
Ichio Ii and Keitaro Tanoi

7 **Effects of "Clean Feeding" Management on Livestock Products Contaminated with Radioactive Cesium Due to the Fukushima Daiichi Nuclear Power Plant Accident** 77
Noboru Manabe, Tomotsugu Takahashi, Maiko Endo, Chunxiang Piao, Junyou Li, Hiroshi Kokado, Minoru Ohta, Keitaro Tanoi, and Tomoko M. Nakanishi

8 Adverse Effects of Radiocesium on the Promotion of Sustainable Circular Agriculture Including Livestock Due to the Fukushima Daiichi Nuclear Power Plant Accident 91
Noboru Manabe, Tomotsugu Takahashi, Chunxiang Piao, Junyou Li, Keitaro Tanoi, and Tomoko M. Nakanishi

9 Wild Boars in Fukushima After the Nuclear Power Plant Accident: Distribution of Radiocesium 99
Keitaro Tanoi

10 Contamination of Wild Animals: Microhabitat Heterogeneity and Ecological Factors of Radioactive Cesium Exposure in Fukushima 107
Ken Ishida

11 Translocation of Radiocesium in Fruit Trees 119
Daisuke Takata

12 The Effects of Radioactive Contamination on the Forestry Industry and Commercial Mushroom-Log Production in Fukushima, Japan 145
Satoru Miura

13 Radiocesium in Timber of Japanese Cedar and Japanese Red Pine, in the Forests of Minamisoma, Fukushima 161
Masaya Masumori, Norio Nogawa, Shin Sugiura, and Takeshi Tange

14 Ecosystem Monitoring of Radiocesium Redistribution Dynamics in a Forested Catchment in Fukushima After the Nuclear Power Plant Accident in March 2011 175
Nobuhito Ohte, Masashi Murakami, Izuki Endo, Mizue Ohashi, Kohei Iseda, Takahiro Suzuki, Tomoki Oda, Norifumi Hotta, Keitaro Tanoi, Natsuko I. Kobayashi, and Nobuyoshi Ishii

15 Reduction of Air Radiation Dose by Ponding Paddy Fields 189
Naritaka Kubo, Toshiaki Iida, and Masaru Mizoguchi

16 Collaboration Structure for the Resurrection of Iitate Village, Fukushima: A Case Study of a Nonprofitable Organization 205
Hanae Yokokawa and Masaru Mizoguchi

17 Impacts of the Nuclear Power Plant Accident and the Start of Trial Operations in Fukushima Fisheries 217
Nobuyuki Yagi

18 Consumer Evaluation of Foods from the Disaster Affected Area: Change in 3 Years 229
Hiromi Hosono, Yuko Kumagai, Mami Iwabuchi, and Tsutomu Sekizaki

19 Imaging Techniques for Radiocesium in Soil and Plants 247
Ryohei Sugita, Atsushi Hirose, Natsuko I. Kobayashi, Keitaro Tanoi, and Tomoko M. Nakanishi

Chapter 1
An Overview of Our Research

Tomoko M. Nakanishi

Abstract The movement of radioactive Cs emitted from the Fukushima Nuclear accident has been studied by the academic staff of the Graduate School of Agricultural and Life Sciences, The University of Tokyo. The targeted items for research ranged widely, including soils, plants, animals, fish, mountains, water, etc. The relevant feature, with regard to the fallout, is that the radioactive Cs has remained at the initial contact sites and has hardly moved since. However, in the case of living individuals, such as animals, the amount of radioactivity has decreased with time at a much faster rate than the physiological half-life because of their metabolic activities. The biological half-life in animals was estimated to be within a period of 100 days. Soil plays a major role in fixing fallout. When fallout nuclides are adsorbed into the soil, plants growing there can absorb little of the radioactive Cs. In the mountains, radioactive Cs was gradually transferred from litter to soil and moved little even when washed with heavy rains. The method of contamination by radioactive nuclides is completely different from that of heavy metals.

Keywords Fukushima nuclear accident • Fallout • Radioactive Cs movement • Soil • Plant • Forest • The method of contamination

1.1 Research Project

Immediately after the Fukushima Nuclear accident, the academic staff at the Graduate School of Agricultural and Life Sciences, The University of Tokyo, organized several groups to research into the behavior of radioactive materials in the Fukushima prefecture. The researchers were divided into the following six large groups and they still continue their work today:

1. Crop plants and soils
2. Livestock and dairy products

T.M. Nakanishi (✉)
Graduate School of Agricultural and Life Sciences, The University of Tokyo, 1-1-1 Yayoi, Bunkyo-ku, Tokyo 113-8657, Japan
e-mail: atomoko@mail.ecc.u-tokyo.ac.jp

T.M. Nakanishi, K. Tanoi (eds.), *Agricultural Implications of the Fukushima Nuclear Accident*, DOI 10.1007/978-4-431-55828-6_1

3. Fishery
4. Environment, including wild life and forest
5. Radioactivity measurement
6. Science communication

Rice is one of the important cereal crops in Japan; however, most of the rice plants growing in the contaminated soil showed very low radioactivity in the grain, $<$500 Bq/kg, which was the initial regulation value. Adding K to soil proved to be an excellent and most effective way of preventing radioactive Cs uptake by plants. However, there were some exceptional cases where radioactivity in the rice grain was $>$100 Bq/kg, which has become the current revised regulation value for foods. Such cases are rare, and they have been studied by Prof. Keisuke Nemoto in an attempt to analyze the origin and chemical form of the radioactive Cs that is easily absorbed by rice plants in a paddy field. Rice is grown and harvested once a year; therefore, only one set of data is being collected per year. Thus, only a few data sets have been accumulated since the accident. This makes it difficult to estimate future contamination of the cereal crop until more data sets are available. In the case of crop radioactivity inspection, the Fukushima prefecture has established a system to measure the radioactivity of all the rice grains before its transfer to market. More than 10 million rice bags, each containing 30 kg of rice grain produced in Fukushima, were measured every year. The contaminated rice grain was not sold.

To measure radioactive Cs, we use pure Ge counters and Na(Tl)I counters in our radioisotope lab. An enormous number of samples of various types, mainly collected by our colleagues, were brought to this lab, and their radioactivity was measured by professional employees. Over 10,000 samples were measured per year.

Within a few months after the Fukushima Nuclear accident, most of the radioactive nuclides measured were radioactive Cs because other radioactive nuclides, such as ^{131}I, had decayed out because of their relatively short half-lives. Figure 1.1 is an example of measurements using a pure Ge counter showing the gamma-ray spectrum of rice grains. The gamma-ray energy used to calculate the nuclides ^{134}Cs, ^{137}Cs, and ^{40}K was 604.7, 661.7, and 1460.8 keV, respectively. The respective detection limits of the nuclides ^{134}Cs, ^{137}Cs, and ^{40}K were 0.7, 0.8, and 23.8 Bq/kg.

In our previous book (Nakanishi and Tanoi 2013), we reported that soil and plant contamination are different from that of animals. The features of these contaminations are summarized briefly below, with new findings that were not previously included.

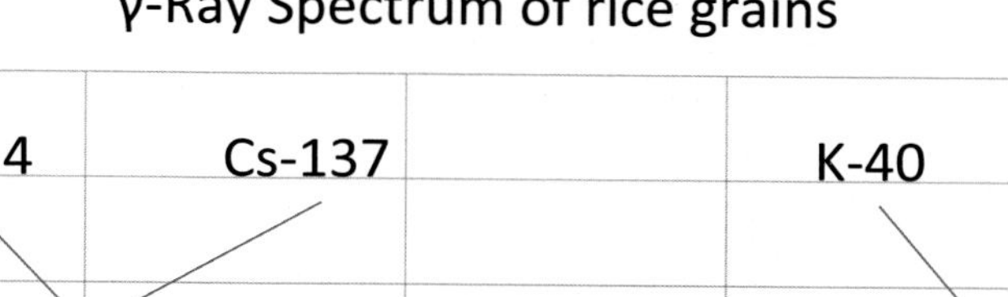

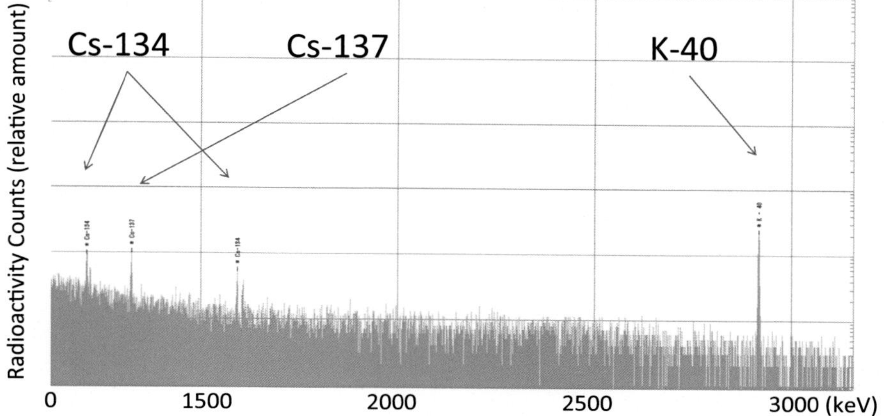

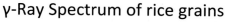

Fig. 1.1 An example of gamma-ray spectrum of rice grains

1.2 Fallout on Soil, Crops, and Trees

1.2.1 Soil

The role of the soil was the most important consideration in the movement of radioactive Cs which is adsorbed into the very fine clay and organic matter. When soil from the contaminated fields was collected and radioactivity images of the fallout were taken, they showed spot-like distribution even after several months. In addition, the fallout showed no movement. The radioactive Cs was difficult to separate from the soil by chemical treatment and could only be washed from the soil during the first few weeks after fallout; the adsorption of radioactive nuclides became stronger with time, thereby making them more difficult to remove.

Prof. Sho Shiozawa has been measuring the radioactivity of the soil along with the depth. He set several pipes vertically in the soil and periodically measured the radioactivity in the pipes along with the depth of the soil. He found that the downward movement of radioactive fallout is now about 1–2 mm/year, whereas in the first 3 months after the accident it moved approximately 20 mm/3 months and then, for the following 3 months, it moved approximately 6 mm/3 months. The speed of the movement is now much slower.

Prof. Shiozawa also measured the radioactivity at the surface of the basement soil of ponds, under the water, using a special waterproof survey meter prepared by himself. The radioactivity at the surface of the soil showed little downward movement with time, similar to land soil. There are two main radioactive nuclides detectable now: ^{137}Cs and ^{134}Cs. In most cases, the radioactivity of soil under water decreased with time, especially because of the relatively short half-life of ^{134}Cs, i.e., 2 years. The ponds he selected were collecting water from the mountains, and he recorded that little radioactive Cs was flowing into the ponds even after heavy rains.

He found only one pond where the radioactivity of the soil surface under the water was not decreased. He examined the location and environment of the pond, especially the surrounding upper mountainous area, and he found that there was a small village in close proximity to the pond where the people were always washing roads, houses, etc. using water. As a result of this activity, it seems that some of the radioactive Cs was removed from the concrete surfaces and flowed into the pond. In a few years, he will summarize and report these findings.

1.2.2 Crops

The Fukushima Nuclear accident occurred in March, and 2 months later, the ears of wheat crops developed and were ready for harvest. When the distribution of the radioactivity in the wheat was measured, to our great surprise, we found that it was more concentrated in the old leaves, which were exposed to the air at the time of the accident. The radioactivity of the leaves or ears that developed after the accident was comparatively very low. The fallout nuclides had hardly moved from the place where they had first landed, even after a few months.

When the radioactivity image on the leaves was magnified, the shape was still spot-like. If the radioactive Cs was incorporated into the leaves and had moved along the phloem or xylem, the vein should have been visible in the leaves. The behavior of the radioactive Cs emitted from the nuclear accident was different from the so-called macroscopic Cs chemistry we know. Because the amount of Cs deposited on leaves was so small, and most of them were carrier-free, the nuclides seem to behave like radio-colloids, or as if they were electronically adsorbed onto the tissue.

The radioactive Cs was adsorbed into the soil; therefore, it was unavailable for plant absorption. The real-time moving pictures taken by Dr. Natsuko I. Kobayashi were very convincing to many people. She grew rice plants both in water culture solution and in paddy soil which had been collected from Fukushima. Subsequently, ^{137}Cs was supplied to both the water culture solution and the paddy soil, and a comparison was made between the plants' absorption of ^{137}Cs. In water culture, the plants absorbed high amounts of ^{137}Cs within hours, and it was possible to trace its progress. Because in water culture, ^{137}Cs dissolved as an ion, it was easy for plants to absorb it. However, in the paddy soil, ^{137}Cs was trapped firmly by the soil and was unavailable to the plants. Using both ^{137}Cs and ^{42}K tracers, Dr. Kobayashi noted the effect of K on Cs uptake and the manner of translocation of Cs in the presence of K.

1.2.3 Trees

In the forests, leaves became highly contaminated at the time of the accident, and after falling to the ground and being decomposed by microorganisms, the radioactive Cs that was initially absorbed into the leaves became available to be absorbed by soil. It was found that radioactive Cs adsorption has been moving from the leaves to the soil in the forest. The radioactivity in the forest is also decreasing, along with the decay of ^{134}Cs, which has a half-life of 2 years.

However, as shown in our previous book, the radioactivity of mushrooms growing in the forest is not decreasing. Some of the mushrooms are still accumulating radioactive Cs that originated from the fallout of the nuclear test bomb during the 1960s.

In the case of fruit trees, the first question concerns the origin of the radioactive Cs in the fruit. It was taken for granted that nutrients are absorbed by roots and delivered to the whole tree. However, the fallout remained at the surface of the soil, away from the roots. Despite the active root of the peach tree being about 30 cm below the surface of the soil, where radioactive nuclides do not exist, the fruits still accumulated radioactive Cs and were slightly contaminated. Few people had considered that radioactive Cs moves into the bark from the trunk surface and could then be transferred to the fruits. Dr. Daisuke Takada performed numerous experiments to determine how radioactive Cs moves within trees, and now some of the new findings are described in this book.

1.2.4 Summary of Soil, Crop, and Tree Contamination

Most of the radioactive Cs remains adsorbed on the surface of the substances it first contacted, and the radioactivity image of radioactive Cs still maintains a spot-like shape, indicating its presence. It is very difficult to remove radioactive Cs from soil or plants (Fig. 1.2).

Below is a summary of the features of the fallout, i.e., information in the previous book:

1. The contamination in soil, crops, and trees were found as spots.
2. Emitted nuclides stay where they first landed.
3. They rarely move and stay as spots even after a few years.
4. Only a small portion of the fallout was dissolved in solution and moved.
5. The adsorption manner became stronger with time.
6. Supplying K as a fertilizer is the most effective and efficient way to prevent Cs uptake by plants.

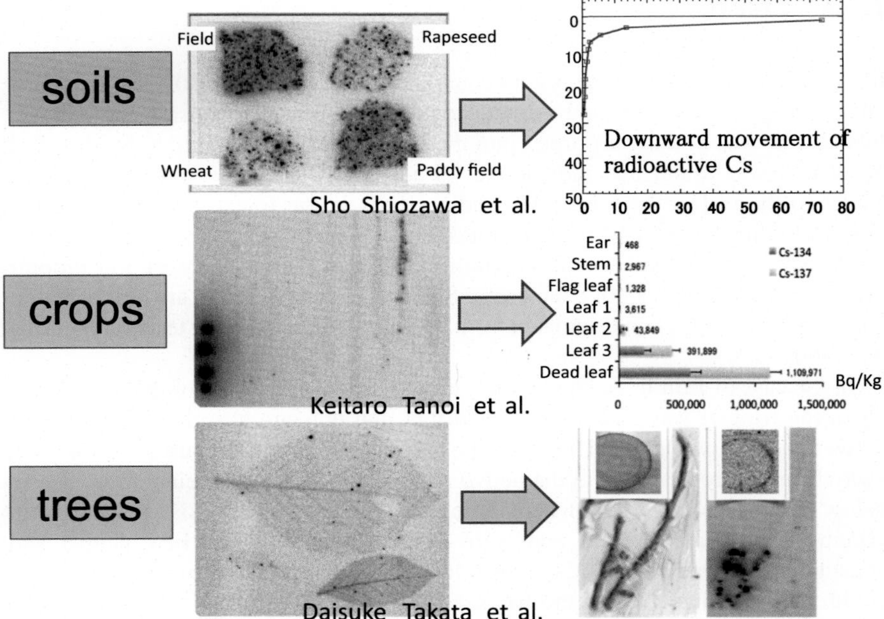

Fig. 1.2 Features of fallout for soil, crops, and trees. Radioactive Cs was adsorbed in spot-like distribution at the surface and rarely moved with time

1.3 Fallout on Birds, Fish, and Livestock

1.3.1 Birds

In the case of birds, it was difficult to identify their point of contamination because they can fly long distances. Associate Prof. Ken Ishida caught several birds in Fukushima and compared the contamination of their feathers to those caught in other areas. A radiograph of the feathers revealed the spot-like distribution of the radioactivity. This distribution pattern was similar to that found in soil or plants. Radioactive Cs was not removed by washing with chemicals. In the following year, the same species of birds were caught and examined; however, no radioactivity was measured in their feathers. The contaminated feathers seemed to have been renewed or replaced as contaminated birds were only found in the year of the accident.

1.3.2 Fish

The study of contaminated fish was conducted mainly with respect to food safety. First of all, Prof. Toyoji Kaneko showed how fish excrete Cs into the water, since they eliminate radioactive Cs faster than its half-life. He showed that fish excrete Cs in the same way as K from their scale cells.

Fish is a major food source in Japan and many products originate from fish meat. Prof. Shugo Watabe studied the safety of food products. He measured how radioactivity was reduced during food processing. He targeted a popular food called Kamaboko, produced from fish meat paste and found <5 % of the radioactivity in the original fish meat remained in the final products. He found that most of the radioactive Cs was removed during the process of washing the fine homogenated meat.

1.3.3 Livestock

In the case of animals, we found that radioactive Cs appeared in milk soon after contaminated feeds were given. However, when non-contaminated foods were supplied, the radioactive materials in the animals were metabolized and decreased. Similar results were found for animal meat, indicating that when contaminated animals are identified, it is possible to decontaminate them by feeding them non-contaminated feeds. In the contaminated area, natural mating of pigs and wild boars take place and the number of hybrid animals is on the increase. While eating, pigs and wild boars habitually dig in the soil and seem to inhale or eat a portion of the surface soil. In comparison, cows eat only plants. As a consequence of this difference in activity, radioactive Cs in the meat of pigs or wild boars is much higher than that of cows.

1.3.4 Summary of Bird, Fish, and Livestock Contamination

In the case of birds, fish, and livestock, radioactive Cs was found on feathers, muscles, and meat or dairy products, respectively. Radioactivity in living animal tissue was rapidly decreased by feeding non-contaminated foods. The biological half-life of ^{137}Cs was estimated to be <100 days because of the biological activity and metabolism, whereas the physical half-life of ^{137}Cs is 30 years. To summarize the findings, the common features of radioactive Cs fallout are shown in Fig. 1.3.

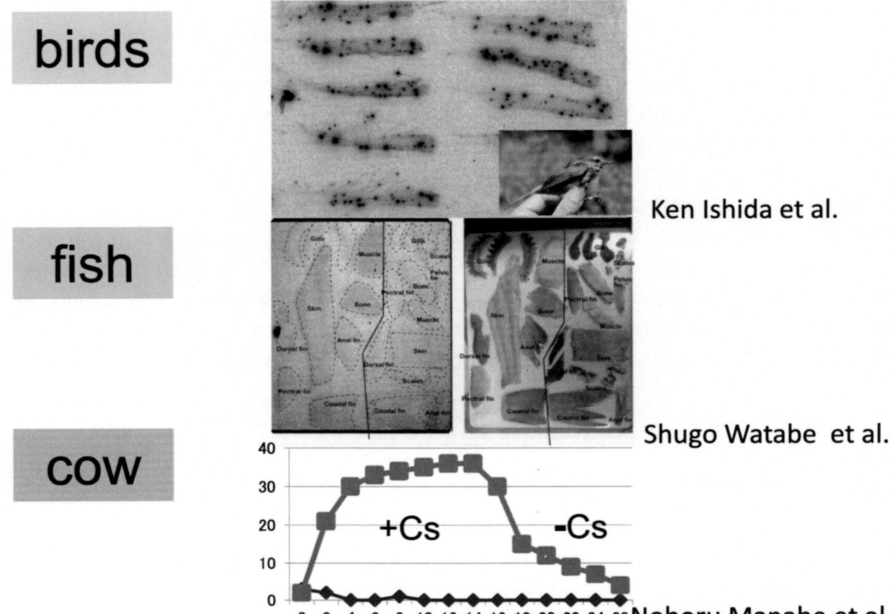

Fig. 1.3 Features of fallout for birds, fish, and livestock. In living individuals, radioactive Cs is decreased through metabolism. The biological half-life of ^{137}Cs is within 100 days

1.4 Radioactive Contamination

There is a great difference between radioactive contamination and contamination from heavy metals. Some 50–60 years ago in Japan, we experienced much heavy metal contamination in the environment causing disease to humans. These contaminants, such as Cd or Hg, became dissolved in water and spread in the environment. Subsequently, the contaminated plants, animals, or water were taken up by people as food and caused serious diseases. For this reason, we fear the movement of contaminants into the environment. In the case of radioactive nuclides, though they were emitted from the nuclear power plant, the radioactive fallout did not spread far away after the settlement. Therefore, it is important to know where the radioactive nuclides reside in the environment and how they move with time as well as what is the most effective way to shield the radiation.

At the same time, we need to understand the features of the fallout, including the types and amount of radioactive nuclides changing with time after the accident.

Because Japan is located in a monsoon area with many paddy fields used to grow rice, the agricultural environment is similar to that in other Asian countries. The climate and agricultural environment in Japan is different from those in Chernobyl; therefore, it is important to gather information regarding the movement or features of fallout specific to Japan from an agricultural point of view.

Reference

Nakanishi TM, Tanoi K (eds) (2013) Agricultural implications of the Fukushima nuclear accident. Springer, Tokyo

Chapter 2
Monitoring Inspection for Radioactive Substances in Agricultural, Livestock, Forest and Fishery Products in Fukushima Prefecture

Naoto Nihei

Abstract We summarize the inspections of radiocesium concentration levels in agricultural, livestock, forest and fishery produced in Fukushima Prefecture, Japan, for 3 years from the nuclear accident in 2011. The ratio in which radiocesium concentration exceeded the 100 Bq/kg from March 2011 to June 2011 was 18 % in agricultural products (excluding rice), 3 % in livestock productions, 49 % in forest productions, and 52 % in fishery produced. After June 2011, radiocesium concentration reduced drastically. Radiocesium concentration in agricultural and livestock products hardly exceeded the 100 Bq/kg. On the other hand radiocesium concentrations of forest and fishery products have been falling every year, but there were a little high concentration samples. Soybean is one of the agricultural products, and the radiocesium concentration is higher than the other agricultural products. We analyzed the absorption process in soybean in contaminated areas. The radiocesium concentration of the above-ground part was lateral root > leaf ≒ petiole ≒ pod > stem ≒ main root. There was a difference in concentration ratio of radiocesium: potassium among parts of the plant. Comparing 10 soybean varieties, radiocesium concentration of wild soya bean showed more than twice as high as other varieties. And the radiocesium inside the soybean grain was distributed generally uniformly throughout the entire grain.

Keywords Monitoring inspection • Agricultural products • Livestock products • Forest products • Fishery products • Radiocesium • Fukushima prefecture

---

N. Nihei (✉)
Graduate School of Agricultural and Life Sciences, The University of Tokyo, 1-1-1, Yayoi, Bunkyo-ku, Tokyo 113-8657, Japan
e-mail: anaoto@mail.ecc.u-tokyo.ac.jp

T.M. Nakanishi, K. Tanoi (eds.), *Agricultural Implications of the Fukushima Nuclear Accident*, DOI 10.1007/978-4-431-55828-6_2

2.1 Introduction

The Great East Japan Earthquake occurred on March 11, 2011, and was immediately followed by the accident at the Fukushima Dai-ichi Nuclear Power Plant (NPP) of Tokyo Electric Power Company (hereafter referred to as the nuclear accident). Radioactive materials released during the accident reached agricultural lands in Fukushima and neighboring prefectures and contaminated the soil and agricultural products (Yasunari et al. 2011; Zheng et al. 2014). To fulfill the requirements of the Food Sanitation Act in Japan (Law No. 233 issued in 1947), on March 17, 2011, the Ministry of Health, Labor and Welfare, Japan (MHLW) established a provisional regulation level of 500 Bq/kg for radiocesium in cereals, vegetables, meat, and fishery products. In April 1, 2012, a new maximum limit of 100 Bq/kg was established as a new standard of radiocesium in general food excluded infant food, milk, water and beverages (Hamada et al. 2012). To revitalize agriculture within the prefecture, Fukushima Prefecture has been promoting the decontamination of agricultural land while implementing radioactive substance absorption suppression measures for agricultural products. To verify the safety of agricultural, livestock, forest, and fishery products, the Nuclear Emergency Response Headquarters have been conducting emergency environmental radiation monitoring of agricultural and fishery products (hereafter referred to as monitoring inspections) as part of the emergency response in accordance with the special measure of the Nuclear Disaster Act in Japan. Targeted items and sampling locations for the monitoring inspections are determined by discussions with municipalities, consideration of the amount of production, and value of any shipment. When the results of a monitoring inspection indicate that the level of radiocesium exceeds 500 Bq/kg (the provisional regulation level, from March 17 in 2011 until March 31 in 2012) or 100 Bq/kg (the new standard, after April 1 in 2012), the relevant municipalities are requested to restrict shipments, based on instructions issued by the Director-General of the Nuclear Emergency Response Headquarters in Japan. This paper describes the polluted situation of the agricultural, livestock, forest and fishery products based on the results of the monitoring for 3 years after the nuclear accident. And soybean is one of the agricultural products, and the radiocesium concentration is higher than the other agricultural products. We analyzed the absorption process in soybean in contaminated areas.

2.2 Radiocesium Concentrations in Agricultural, Livestock, Forest and Fishery Products for Three Years After the Nuclear Accident in Fukushima Prefecture

The investigative program required almost 1 week for each item: agricultural, livestock, forest and fishery products. The extracted samples were washed with tap water and the edible portions were chopped finely. The samples were packed in

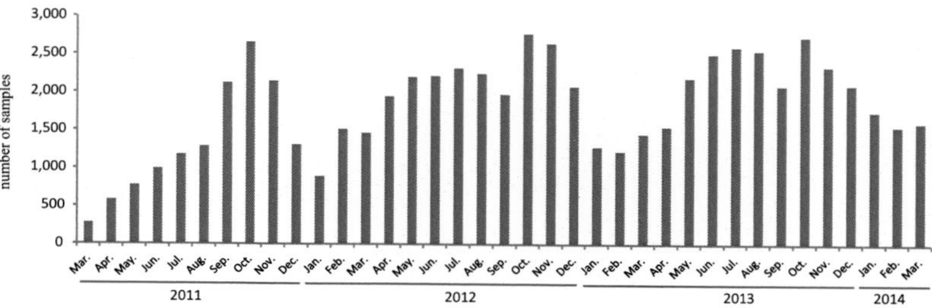

Fig. 2.1 Monthly number of samples

a container and were measured using the germanium semiconductor detector (CANBERRA) at Fukushima Agricultural Technology Centre. The detection limit for radiocesium was approximately 10 Bq/kg. Approximately 500 items, were monitored to produce 67,000 data points by the end of March 2014 (Fig. 2.1). We considered the samples that were harvested from March to June 2011, from July to March 2012, from April 2012 to March 2013, from April 2013 to March 2014 (Table 2.1). The radiocesium concentration in each items were classified as below 10 Bq/kg, from 10 to 100 Bq/kg, from 100 to 500 Bq/kg, and >500 Bq/kg. The results of the monitoring have been released on the homepages of Fukushima Prefecture and the Ministry of Health, Labour, and Welfare, Japan.

Fukushima Prefecture: http://www.new-fukushima.jp/monitoring.php.

Ministry of Health, Labour, and Welfare: http://www.mhlw.go.jp/stf/houdou/2r9852000001m9tl.html.

2.2.1 *Agricultural Products*

Agricultural products include cereals (soybeans, adzuki beans, buckwheat, wheat, etc., rice is described at a different chapter.), vegetables (spinach, cucumbers, carrots, tomato, etc.), and fruits (peaches, apples, pears, etc.). Figure 2.2 shows the concentration of radiocesium and radioiodine in the samples harvested from March 2011 to March 2014. Concentrations of radiocesium in agricultural products were the highest immediately after the nuclear accident and rapidly decreased within the first 3 months. The ratio in which radiocesium concentration exceeded the 100 Bq/kg from March 2011 to June 2011 was 18 %, and the maximum value was 82,000 Bq/kg. After July 2011, these concentrations decreased sharply. The ratio in which radiocesium concentration exceeded the 100 Bq/kg was 3 % from July 2011 to March 2012, 0 % from April 2012 to March 2013, and 1 % from April 2013 to March 2014. Higher radiocesium concentrations in agricultural products were observed in soybean, wasabi, plums, kiwis. Radioactive pollution of agricultural products could be divided into direct pollution, in which substances are deposited directly onto the agricultural products, and indirect pollution, in which

Table 2.1 Radiocesium concentration after Fukushima DNPP accident

Period			Agricultural products	Forest products	Livestock products	Fishery products
3/17/2011~6/30/2011	Number of samples		1496	430	387	321
	<10 Bq/kg	%	66	36	85	14
	10~100 Bq/kg	%	16	16	12	34
	100~500 Bq/kg	%	10	26	3	36
	500 Bq/kg<	%	8	23	0	16
	Max	Bq/kg	82,000	13,000	510	14,400
7/1/2011~3/31/2012	Number of samples		5186	653	5501	3236
	<10 Bq/kg	%	79	56	90	24
	10~100 Bq/kg	%	17	30	9	41
	100~500 Bq/kg	%	3	9	0	27
	500 Bq/kg<	%	0	5	0	5
	Max	Bq/kg	2400	28,000	460	18,700
4/1/2012~3/31/2013	Number of samples		9450	1180	6895	6895
	<10 Bq/kg	%	90	60	99	46
	10~100 Bq/kg	%	9	32	1	41
	100~500 Bq/kg	%	0	6	0	12
	500 Bq/kg<	%	0	2	0	2
	Max	Bq/kg	1460	5600	146	1004
4/1/2013~3/31/2014	Number of samples		10,378	1465	5476	8497
	<10 Bq/kg	%	85	69	99	72
	10~100 Bq/kg	%	14	25	1	25
	100~500 Bq/kg	%	1	4	0	3
	500 Bq/kg<	%	0	2	0	0
	Max	Bq/kg	342	11,870	83	1720

agricultural products absorb substances from the soil through their roots. Direct pollution has a greater impact than indirect pollution. The data after July 2011 inspected mainly the agricultural products that had been grown after the nuclear accident, and the impact of the radiocesium was mainly indirect pollution. Therefore, these concentrations decreased sharply. A little samples showed a higher radiocesium concentration than 100 Bq/kg after April 2012, this might be because these samples were grown with using the agricultural materials which were left outside during the nuclear accident.

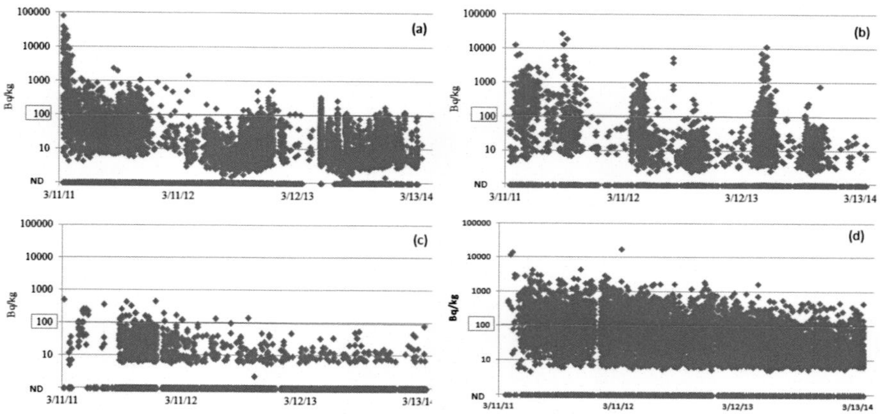

Fig. 2.2 Progress of radiocesium concentration after Fukushima DNPP accident. (**a**) agricultural products, (**b**) forest products, (**c**) livestock products, (**d**) fishery products, *ND* not detected, 100 Bq/kg: the new standard value of radiocesium concentration

2.2.2 Radiocesium Absorption in Soybean

Soybean is the global production volume is approximately 250 million tons. This is the fourth largest production volume after wheat, rice and corn. According to monitoring inspection, the ratio of samples exceeding the new standard value of radiocesium (100 Bq/kg) was found to be 5.7 % for soybean, 2.6 % for rice, 11 % for wheat in 2011, 2.6 % for soybean, 0.0007 % for rice and 0 % for wheat in 2012, 1.9 % for soybean 0.0003 % for rice, 0 % for wheat in 2013 (Fig. 2.3). In other words, the monitoring inspections indicated that the ratio of soybean that had exceeded 100 Bq/kg was high compared with rice and wheat for 3 years and that the tendency for decline was low compared with rice and wheat. There are cultivation areas of soybean following rice. In order to make recovery and revitalization of agricultural industries, analysis of the absorption process in soybean was conducted by cultivation in contaminated areas.

2.2.2.1 Absorption of Radiocesium in Soybean in the Contaminated Areas

To study the absorption process of soybean in the contaminated areas, soybean (var. Fukuibuki) was grown by no fertilizing in Iitate Village, Fukushima Prefecture on June 29, 2013 (Fig. 2.4). The radiocesium of the field was approximately 13,000 Bq/kg (depth 15 cm), exchangeable potassium was 15.8 mg/100 g, and pH was 6.2.

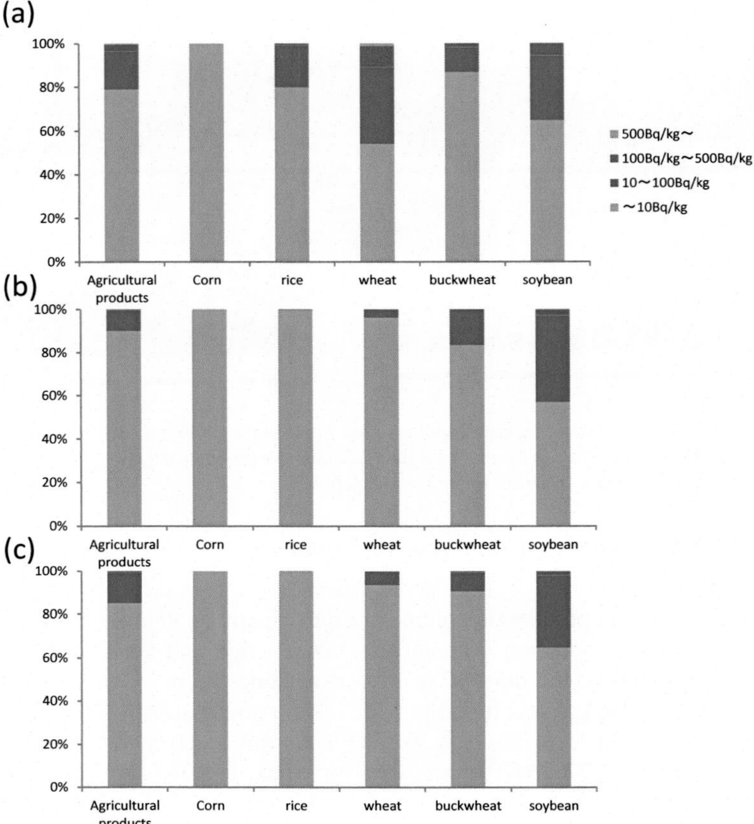

Fig. 2.3 The ratio of radiocesium concentration in the monitoring inspection for 3 years (**a**) 2011, (**b**) 2012, (**c**) 2013

The radiocesium absorption in soybean was observed from the initial stage of growth, and the absorption increased together with the weight of the above-ground portion until the middle of September (maximum growth period) (Fig. 2.5). Soybean continues nutrient growth even after the flowering period, and the period of continuing nutrient growth is thought to be when radiocesium is being absorbed. When the concentration of radiocesium was measured for each part of the plant at the middle of September, the differences were observed. The radiocesium concentration of the above-ground part was observed: lateral root > leaf≒petiole≒pod > stem≒main root (Fig. 2.6). For the concentration of potassium, which is a homologous element of cesium, the following was observed: petiole > pod > leaf≒stem > main root≒lateral root. There was a difference in concentration ratio of radiocesium to potassium. The difference in distribution within the plant between radiocesium and potassium has

Fig. 2.4 Soybean in Iitate Village

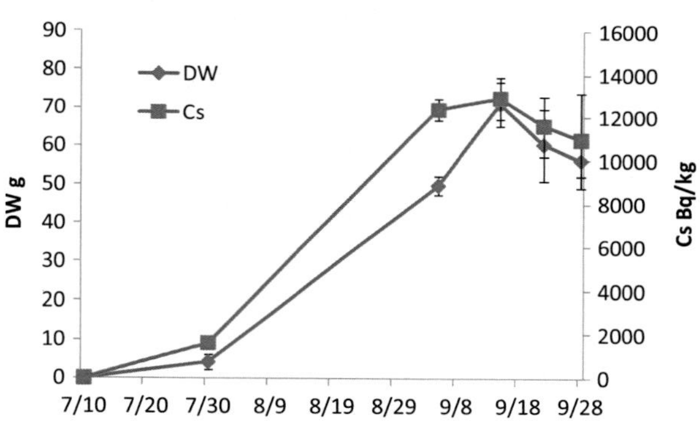

Fig. 2.5 The progress of radiocesium absorption and dry weight in soybean

been pointed out for rice (Tsukada et al. 2002), and the distribution of soybean is similar to rice.

2.2.2.2 Differences Between Soybean Variety

If there is a difference in radiocesium absorption between varieties, it would offer the reduction technology of radiocesium absorption that could immediately be introduced in the contaminated areas, and it would be expected that varietal difference leads also to the mechanism elucidation of cesium absorption. The differences in radiocesium absorption between varieties have been conducted on rice (Ohmori et al. 2014). In this paper, the absorption of radiocesium was studied comparing 10 varieties of soybean in Iitate Village, Fukushima Prefecture. The samples were parental lines that held cross-fertilization later generations, varieties A (wild soya bean) and B which were obtained from the National BioResource Project, and cultivation varieties which had grown inside Fukushima Prefecture (C to J).

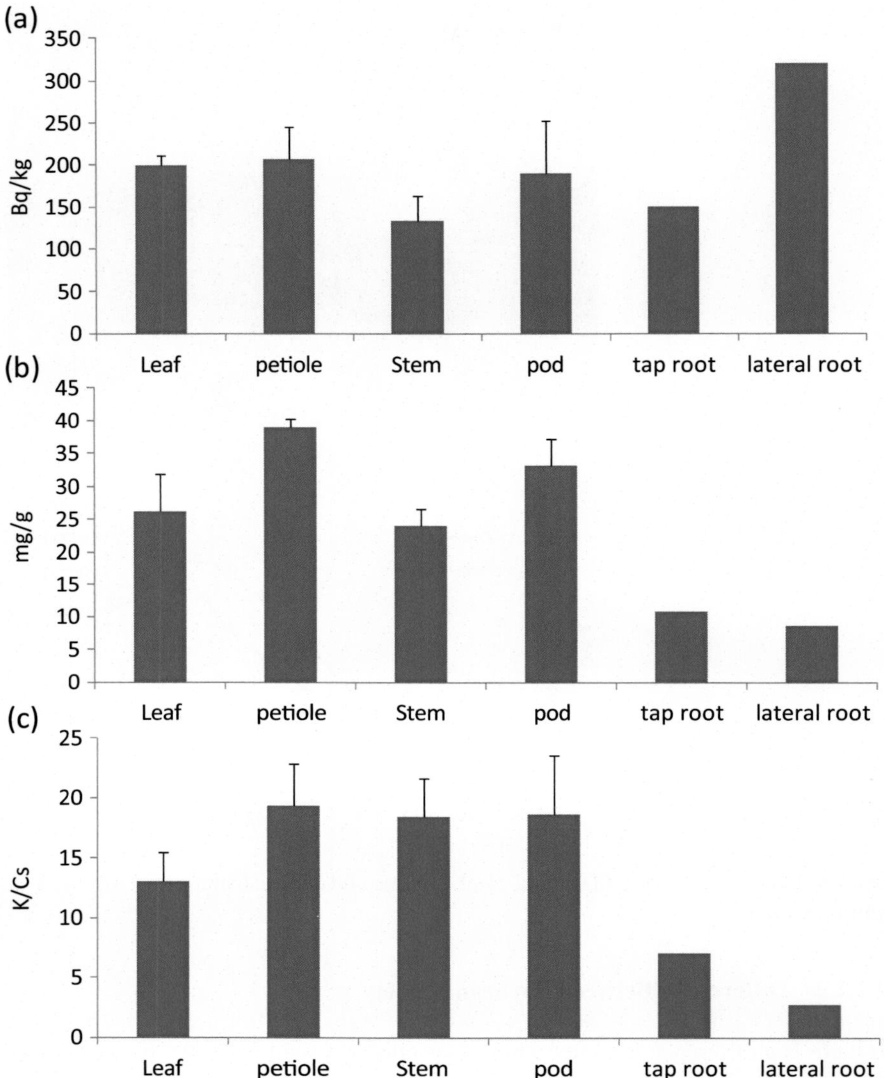

Fig. 2.6 The radiocesium and potassium concentration of soybean (22 Sept 2013). (**a**) radiocesium concentration, (**b**) potassium concentration, (**c**) K/Cs

Figure 2.7 shows the transfer factor to grain of different varieties. When compared with the parental line A and B, A showed a tendency to be about twice as high as B. The transfer factor of the C to J did not have a large difference. No clear tendency based on the grain size was observed. No clear correlation between the grain radiocesium concentration and the grain potassium concentration was

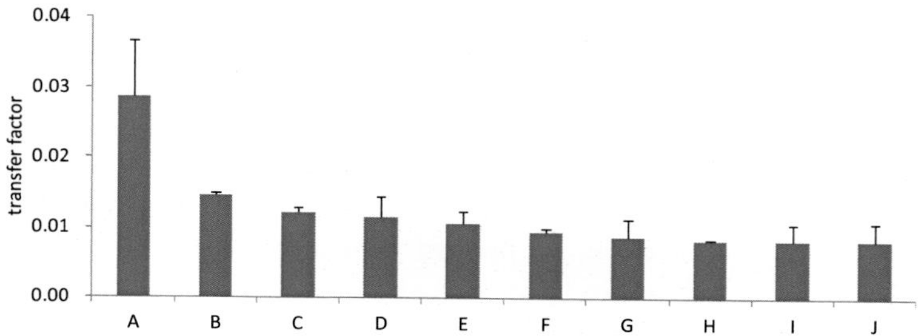

Fig. 2.7 Transfer factor to soybean grains of different varieties

observed. We plan to use later generations of cross-fertilization between A and B to study the genetic locus involved in the absorption of radiocesium.

2.2.2.3 Distribution of Cesium in Soybean Grain

To study the distribution of radiocesium in soybean grain, sections of soybean grain cultivated using liquid culture medium with added radiocesium (^{137}Cs) were contacted to an IP (imaging plate, Fujifilm Co.), and autoradiography was obtained. The radiocesium inside the soybean grain was distributed generally uniformly throughout the entire grain. For the radiocesium of the rice grain, it was reported that radiocesium was topically higher in the embryo and aleurone layer, and we have shown that radiocesium distributions inside soybean and rice are different. This is considered to be due to the difference in seed storage tissue. The rice grain stores nutrients (include potassium) in the embryo, and starch and protein is accumulated in the endosperm. We guess the radiocesium stores in the embryo rather than in the central part of the rice grain. However, as the soybean grain is an exalbuminous seed, the nutrients, starch and lipids accumulate in the cotyledon, which comprises a large portion of the seed. Thus, the nutrients store uniformly inside the grain, and it is thought that the cesium, like the other nutrients, is distributed uniformly inside the grain. The form in which nutrients accumulate is different between rice and soybean, and the mineral constituent of soybean is about five times that of rice. The high accumulation of nutrients inside the grain is thought to be one factor for the high accumulation of cesium in soybean.

2.3 Livestock Products, Forest Products, Fishery Products

Livestock products include beef, pork, raw milk, chicken eggs et al. Monitoring of radioactive concentrations in livestock products began with raw milk immediately after the nuclear accident. Radioactiveiodine that exceeded the provisional regulation level (2000 Bq/kg) was detected in raw milk between March 2011 and June

2011 (data not shown). Beef which were over 500 Bq/kg of radiocesium concentrations were detected between March 2011 and June 2011 (Fig. 2.2). But both radioactive iodine and cesium of livestock products decreased from July 2011. Beef which were over 500 Bq/kg of radiocesium concentrations might have been due to the supply animals with the polluted feeds by radiocesium. The government and Fukushima prefecture have called farmers not to feed hays and grass that was outside at the nuclear accident to prevent further contamination of livestock products.

Forest products include edible wild plants (bamboo shoots, fatsia sprouts, etc.), mushrooms. Measurements for cesium in forestry products began in late March 2011, which is also the beginning of the harvest season for edible wild plants. The ratio in which radiocesium concentration exceeded the 100 Bq/kg was 49 % from March 2011 to June 2011, 14 % from July 2011 to March 2012, 8 % from April 2012 to March 2013, and 6 % from April 2013 to March 2014 (Fig. 2.2). This result identified a high trend in radiocesium concentrations for forestry products compared to agricultural or livestock products. Radiocesium concentrations remained high even after July 2011, suggesting that forestry products were not only polluted directly with radiocesium released from the nuclear disaster, but also indirectly with the absorbed radiocesium from the soil. Especially bamboo shoots, '*koshiabura*' (*Acanthopanax sciadophylloides*), fatsia sprouts (*Aralia elata* Seem.), and wild mushrooms were detected with high radiocesium concentrations.

The ratio of fishery products in which radiocesium concentration exceeded the 100 Bq/kg was 52 % from March 2011 to June 2011, 32 % from July 2011 to March 2012, 14 % from April 2012 to March 2013, and 3 % from April 2013 to March 2014, and the maximum value was 18,700 Bq/kg (Fig. 2.2). This result indicates that radiocesium concentration of fishery products have been falling every year, but high concentration remain as well as forestry products. There are difference according the kind of fish, octopus and squids are low, on the other hand cod and stone flounder are high.

Four years have passed since the accident, and we expect to see the sequential reopening of agriculture even in the evacuated regions. In order to provide safe agricultural crops, not only do we think it is necessary to continue monitoring tests, but we would also like to continue research that contributes to the efforts of agricultural recovery and revitalization in the areas contaminated by the nuclear accident.

References

Hamada N, Ogino H, Fujimichi Y (2012) Safety regulations of food and water implemented in the first year following the Fukushima nuclear accident. Radiat Res 53:641

Ohmori Y et al (2014) Difference in cesium accumulation among rice cultivars grown in the paddy field in Fukushima prefecture in 2011 and 2012. J Plant Res 127:57–63

Tsukada H, Hasegawa H, Hisamatsu S, Yamasaki S (2002) Rice uptake and distribution of radioactive ^{137}Cs, stable ^{133}Cs and K from soil. Environ Pollut 117:403–409

Yasunari TJ, Stohl A, Hayano RS, Burkhart JF, Eckhardt S, Yasunari T (2011) Cesium-137 deposition and contamination of Japanese soils due to the Fukushima nuclear accident. Proc Natl Acad Sci USA 108:19530

Zheng J, Tagami K, Bu W, Uchida S, Watanabe Y, Kubota Y, Fuma S, Ihara S (2014) 135Cs/137Cs isotopic ratio as a new tracer of radiocesium released from the Fukushima nuclear accident. Environ Sci Technol 48:5433

Chapter 3
Rice Inspections in Fukushima Prefecture

Naoto Nihei

Abstract We summarize the inspections of radiocesium concentration levels in rice produced in Fukushima Prefecture, Japan, for 3 years from the nuclear accident in 2011. In 2011, three types of verifications, preliminary survey, main inspection, and emergency survey, revealed that rice with radiocesium concentration levels over 500 Bq/kg (the provisional regulation level until March 2012 in Japan) was identified in the areas north and west of the Fukushima nuclear power plant. The internal exposure of an average adult eating rice grown in the area north of the nuclear plant was estimated as 0.05 mSv/year. In 2012, Fukushima Prefecture authorities decided to investigate the radiocesium concentration levels in all rice using custom-made belt conveyor testers. Notably, rice with radiocesium concentration levels over 100 Bq/kg (the new standard since April 2012 in Japan) were detected in only 71, 28 and 2 bags out of the total 10,338,000 in 2012, 11,001,000 in 2013 and 10,988,824 in 2014, respectively. We considered that there were almost no rice exceeding 100 Bq/kg produced in Fukushima Prefecture after 3 years from the nuclear accident, and the safety of Fukushima's rice were ensured because of the investigation of all rice.

Keywords Rice • Radiocesium • Monitoring inspection • Inspection of all rice • Belt conveyor tester • Fukushima prefecture

3.1 Introduction

The Great East Japan Earthquake occurred on March 11, 2011, and was immediately followed by the accident at the Fukushima Dai-ichi Nuclear Power Plant (NPP) of Tokyo Electric Power Company (hereafter referred to as the nuclear accident). Radioactive materials released during the accident reached farmlands in Fukushima and neighboring prefectures and contaminated the soil and agricultural products (Yasunari et al. 2011; Zheng et al. 2014). Rice is the main staple food of the Japanese diet, and it is the most valuable agricultural product in Fukushima

N. Nihei (✉)
Graduate School of Agricultural and Life Sciences, The University of Tokyo, 1-1-1, Yayoi, Bunkyo-ku, Tokyo 113-8657, Japan
e-mail: anaoto@mail.ecc.u-tokyo.ac.jp

T.M. Nakanishi, K. Tanoi (eds.), *Agricultural Implications of the Fukushima Nuclear Accident*, DOI 10.1007/978-4-431-55828-6_3

Prefecture. Hence, the Central Government of Japan requested suspension of rice planting as a precaution in 2011, based on a special measure of the Nuclear Disaster Act, to the relevant municipal governments that were in the controlled areas within a 20-km radius of the Fukushima Dai-ichi NPP, and where radiocesium exceeding 5000 Bq/kg was detected in the soil. In 2012, as a precaution, rice planting was suspended in areas that produced rice with radiocesium levels exceeding 500 Bq/kg in 2011 and in areas where evacuation orders had been issued. Fukushima Prefecture and the Central Government of Japan have issued more detailed directives for the inspection of rice within the prefecture compared to other agricultural products (Table 3.1). The results of these inspections are publicly announced first in newspapers and then on Fukushima Prefecture webpage and Fukushima-no Megumi Anzen Taisaku Kyogikai webpage. Based on these results, this study analyzes the radioactive content in rice in the 3-year period after the nuclear accident, and it also evaluates the performance of the equipment used to inspect all rice.

3.2 Inspections in 2011

3.2.1 Inspection Method

Monitoring inspections are conducted on three samples in each municipality for each item; however, for rice, the number of samples for inspection is significantly higher. At first, a preliminary survey before harvest was conducted to understand the trends in radioactive content in rice. The sampling was performed about 1 week before harvesting, and a total of 441 samples were analysed. Next, a main inspection was conducted based on the preliminary survey. Two samples were collected for every 15 ha in municipalities where the radiocesium concentration levels exceeded 200 Bq/kg in the preliminary survey, and two samples were collected in every former municipality in other survey sectors. Unpolished rice was collected by reaping from the standing crop and then threshed, dried, and processed. In this manner, a total of 1174 samples were analysed. The inspection method involved filling 100 ml containers with rice and taking measurements for 2000 s using a germanium semiconductor detector at the Fukushima Agricultural Technology

Table 3.1 Inspections of rice from 2011 to 2013

Year	2011	2012	2013
Name of inspection	Preliminary surveys and main inspections	Comprehensive rice bag inspections	Comprehensive rice bag inspections
Measuring instrument	Germanium semiconductor detector	Belt-conveyor-type radiocesium concentration tester	Belt-conveyor-type radiocesium concentration tester
Number of inspection	1624	10,331,526	10,949,026

Centre (conducted in accordance with the "Manual for Measuring Radioactivity of Foods in Cases of Emergency" published by the Ministry of Health, Labour and Welfare, JAPAN (MHLW)). The detection limit was approximately 10 Bq/kg, and the results have been publicly announced on Fukushima Prefecture website. An emergency survey after the main inspection was undertaken by Fukushima Prefecture itself to reassure the public about food safety. The inspections were conducted on one or more samples per farm household in areas where measurements during the main inspection or preliminary survey exceeded detection limit values, 10 Bq/kg. Sampling was performed in a total of 23,247 farm households, using NaI or other types of scintillation counters. The survey methods and results have been publicly announced on the Fukushima Prefecture website.

Preliminary surveys: http://www.pref.fukushima.lg.jp/sec/36035b/23yobichousa-kekka.html

Main inspections: http://www.pref.fukushima.lg.jp/sec/36035b/23honchousa-kekka.html

Emergency surveys: http://www.pref.fukushima.lg.jp/sec/36035b/daishinsai-23komehoushaseibusshitsu-kinkyuuchosa-kekka-syukkaseigen.html

3.2.2 Results

Table 3.2 shows a ratio of radiocesium for every area. For compiling the results, areas were classified into seven administrative sectors according to their distance from the nuclear power plant (Fig. 3.1). In the preliminary survey and main inspection of 2011, 1.7 % and 4.4 % of rice samples had radiocesium concentration levels of 100 Bq/kg or more in Area 1 and Area 3 respectively, which indicates that highly contaminated rice was produced in some parts of Areas 1 and 3. These region were within 100 km northwest of the nuclear power plant, and highly contaminated by the deposited radiocesium (Hirose 2012; Kinoshita et al. 2011), because the plume released from the nuclear power plant from about 12 to 15 JST (Japanese Standard Time) on 15 March 2011 flowed northwestward and wet deposition with precipitation occurred in the nighttime of the same day (Chino et al. 2011). Therefore, the proportion of rice with 100 Bq/kg or higher was greater in these areas than for other areas in 2011. However, there were many samples with 25 Bq/kg or lower in these same areas in 2011; thus, there was a range of radiocesium concentration levels within rice in a given area. Hence, the actual spread of radiocesium was heterogeneous, and the exchangeable potassium content of the soil (Tensho et al. 1961) and the soil type (Tsumura et al. 1984), both of which affect the absorption rate of cesium by crops, were also heterogeneous. In addition, the radiocesium concentration levels in rice were lower in Area 2 (which was at the similar distance from the nuclear power plant as Area 1) and in Areas 4 and 5 (which were at the similar distance from the plant as Area 3) and received a lower radiocesium concentration in the agricultural land. The radiocesium

Table 3.2 Inspection result of rice for 3 years

(a) 2011 (Preliminary surveys and main inspections)

Region	Total number of inspection	Ratio (%)			
		~25 Bq/kg	25 Bq/kg ~ 100 Bq/kg	100 Bq/kg ~ 500 Bq/kg	500 Bq/kg~
Area1	45	40.0	55.6	4.4	0.0
Area2	122	70.5	28.7	0.8	0.0
Area3	576	65.3	33.0	1.7	0.0
Area4	370	89.5	10.5	0.0	0.0
Area5	155	87.1	12.9	0.0	0.0
Area6	306	98.7	1.3	0.0	0.0
Area7	50	100.0	0.0	0.0	0.0
Total	1624	79.9	19.3	0.8	0.0

(b) 2012 (Comprehensive rice bag inspections)

Region	Total number of inspection	Ratio (%)			
		~25 Bq/kg	25 Bq/kg ~ 100 Bq/kg	100 Bq/kg ~ 500 Bq/kg	500 Bq/kg~
Area1	204,315	99.6	0.4	0.0	0.0
Area2	519,593	99.7	0.3	0.0	0.0
Area3	1,299,453	99.0	1.0	0.0	0.0
Area4	3,328,643	99.9	0.1	0.0	0.0
Area5	1,520,043	99.8	0.2	0.0	0.0
Area6	3,153,887	100.0	0.0	0.0	0.0
Area7	305,592	100.0	0.0	0.0	0.0
Total	10,331,526	99.8	0.2	0.0007	0.0

(c) 2013 (Comprehensive rice bag inspections)

Region	Total number of inspection	Ratio of radiocesium (%)			
		~25 Bq/kg	25 Bq/kg ~ 100 Bq/kg	100 Bq/kg ~ 500 Bq/kg	500 Bq/kg~
Area1	259,172	98.8	1.2	0.0	0.0
Area2	558,018	100.0	0.0	0.0	0.0
Area3	1,388,313	99.8	0.2	0.0	0.0
Area4	3,517,451	100.0	0.0	0.0	0.0
Area5	1,582,008	100.0	0.0	0.0	0.0
Area6	3,330,114	100.0	0.0	0.0	0.0
Area7	313,950	100.0	0.0	0.0	0.0
Total	10,949,026	99.9	0.1	0.0003	0.0

(d) 2014 (Comprehensive rice bag inspections)

Region	Total number of inspection	Ratio of radiocesium (%)			
		~25 Bq/kg	25 Bq/kg ~ 100 Bq/kg	100 Bq/kg ~ 500 Bq/kg	500 Bq/kg~
Area1	301,322	99.9	0.1	0.0	0.0
Area2	565,800	100.0	0.0	0.0	0.0
Area3	1,440,598	99.9	0.1	0.0	0.0
Area4	3,575,402	100.0	0.0	0.0	0.0

(continued)

Table 3.2 (continued)

(d) 2014 (Comprehensive rice bag inspections)					
	Total number of inspection	Ratio of radiocesium (%)			
Region		~25 Bq/kg	25 Bq/kg ~ 100 Bq/kg	100 Bq/kg ~ 500 Bq/kg	500 Bq/kg~
Area5	1,548,140	100.0	0.0	0.0	0.0
Area6	3,251,179	100.0	0.0	0.0	0.0
Area7	306,383	100.0	0.0	0.0	0.0
Total	10,988,824	100.0	0.02	0.00002	0.00

concentration level in the agricultural land was also low in Areas 6 and 7, which were more than 100 km away from the nuclear power plant to the west. The proportion of rice with radiocesium content of 25 Bq/kg or lower in Areas 6 and 7 was 98.7 % and 100 %, respectively, indicating minimal impact from radiocesium.

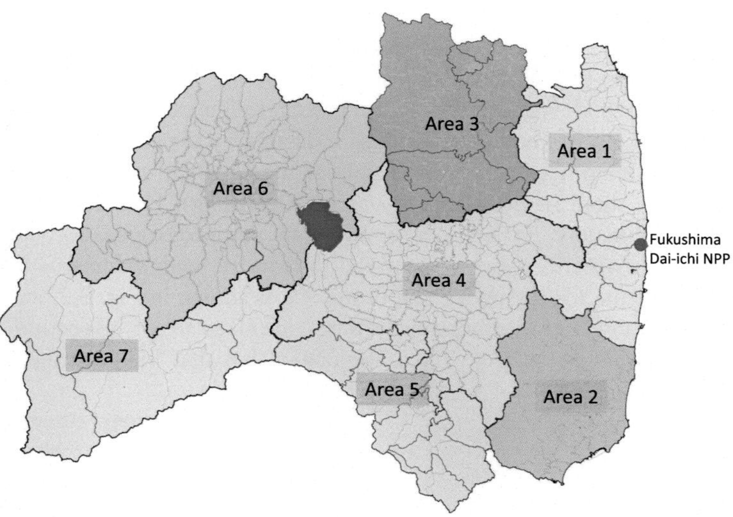

Fig. 3.1 Seven administrative sectors of Fukushima Prefecture. (**a**) Seven administrative sectors of Fukushima prefecture; Area 1 is 30–50 km to the NPP (Soso District), Area 2 is 30–50 km to the south of the NPP (Iwaki District), Area 3 is approximately 30–80 km to the northwest of the NPP (Ken-poku District), Area 4 is approximately 20–70 km west of the NPP (Ken-chu District), Area 5 is approximately 40–80 km to the southwest of the NPP (Ken-nan District), Area 6 is approximately 70–130 km to the west of the NPP (Aizu District), and Area 7 is between 100 and 150 km west of the NPP (Southern Aizu District). Area 1, 2 are called Hamadori (Coastal region); Area 3, 4, and 5 are called Nakadori (Central region); and Area 5, 6 are called Aizu. *NPP* nuclear power plant

3.3 Inspections in 2012, 2013 and 2014

3.3.1 Inspection Method

The inspections conducted by Fukushima Prefecture in 2012 targeted all the rice produced within Fukushima Prefecture (approximately 360,000 t) to reassure the public about food safety (named as inspection of all rice in all rice bags, hereafter referred to as inspection of all rice). The measurement of radiocesium concentration of all rice was considered to be the very first challenge in the world.

For this purpose, manufacturers have developed and produced equipment that can efficiently inspect all rice in Fukushima. Moreover, prefectures and municipalities have compiled information on individual farm households and built inspection frameworks.

Because there were limitations on the number of germanium semiconductor detectors available, and monitoring inspections would have taken considerable time, manufacturers were requested to develop a belt-conveyor-type radiocesium concentration tester (hereafter referred to as the belt conveyor tester) for taking measurements. The belt conveyor testers were equipped with NaI or other types of scintillation counters, and the entire measurement section was shielded by lead or iron. Rice bags weighing 30 kg passed along the belt at a rate of two or three rice bags per minute and were examined to ensure whether radiocesium concentration level exceeded 100 Bq/kg, which was stipulated by the Food Sanitation Act. This measurement method was conducted according to the "Screening Method for Radioactive Cesium in Food Products," as indicated by the MHLW, which stipulates that the value of each screening level calculated using individual equipment must be half or more of the standard value (100 Bq/kg). Fukushima Prefecture installed approximately 200 belt conveyor testers in various areas throughout the prefecture, and inspections were performed to coincide with shipments from producers. The scheme for the inspection of all rice is shown in Fig. 3.2, and can be described as follows: (1) farmers carry their rice bags to an inspection station, (2) their rice bags are sealed with a bar code label that includes the farmer's

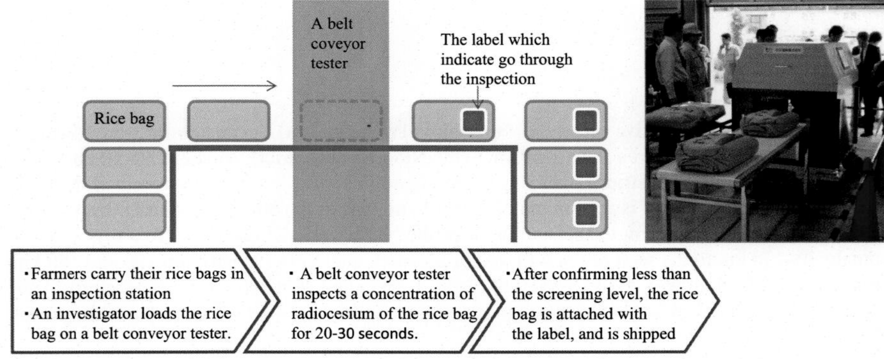

Fig. 3.2 Overview of the inspection of all rice performed from 2012

information, (3) the investigators load the rice bag on a belt conveyor tester and upload the farmer's information with a bar code reader, and (4) the belt conveyor tester measures the radiocesium concentration levels from each rice bag for 20–30 s. If the inspection result is less than the screening level, a label bearing an individual identification number is attached to the rice bag indicating the bag was inspected, and then the bag is shipped. If the screening level is exceeded, the bag is further subjected to a more detailed inspection using a germanium semiconductor detector, and it is isolated and stored until the measurement value is finalized. The result for each rice bag and the results for the rice produced from each area are posted on Fukushima-no Megumi Anzen Taisaku Kyogikai website.

Fukushima-no Megumi Anzen Taisaku Kyogikai (inspection of all rice): https://fukumegu.org/ok/kome/.

3.3.2 Results

The results of the inspection of all rice in 2012 are shown in Table 3.2. For Fukushima Prefecture as a whole (n = 10,338,291), 99.8 % of the locations had measurements of 25 Bq/kg or lower, 0.2 % had measurements higher than 25–100 Bq/kg, and only 71 bags (less than 0.001 %) exceeded 100 Bq/kg. The results of the inspection of all rice in 2013 are similar to that in 2012. Only 28 bags out of 11,001,000 exceeded 100 Bq/kg, which corresponds to 0.0003 %. 99.9 % of rice bags had measurements of 25 Bq/kg or lower.

A comparison of the results from 2011 to 2014 reveals that 0.8 % of all areas in the entire Fukushima Prefecture had contaminated rice with radiocesium concentration levels higher than 100 Bq/kg in the preliminary survey and main inspection in 2011, subsequently only 0.0007 %, 0.0003 % and 0.00002 % of bags had contaminated rice with over 100 Bq/kg of radiocesium in 2012, 2013 and 2014 (Fig. 3.3). The reason for this significant decline is assumed to be the physical reduction of radiocesium in the soil caused by factors such as decay of ^{134}Cs, fixation to the soil clay, decontamination by reversal tillage, and the effects of a thorough soil improvement effort implemented to increase the exchangeable potassium content to approximately 25 mg/100 g (dry soil) or higher, which is the guideline announced by Ministry of Agriculture, Forestry and Fisheries, Japan from 2012. Moreover, planting was restricted in 2012 (by orders from the Central Government of Japan) in areas where rice in 2011 had levels exceeding 500 Bq/kg, most of which was located in Area 3, as shown in Table 3.2.

The inspections were conducted immediately after the harvesting of rice in 2012, 2013, and 99 % of the entire amount was inspected during a 4-month period, from September to December 2012 (Fig. 3.4). In October of 2012 and 2013, about 6,500,000 samples were inspected in a month, which was the peak of the inspection number. Since the inspections were conducted using approximately 200 counters, the inspection in October was performed about 1000 samples per day by a single unit. This means that if measurements are assumed to have been taken over an 8-h

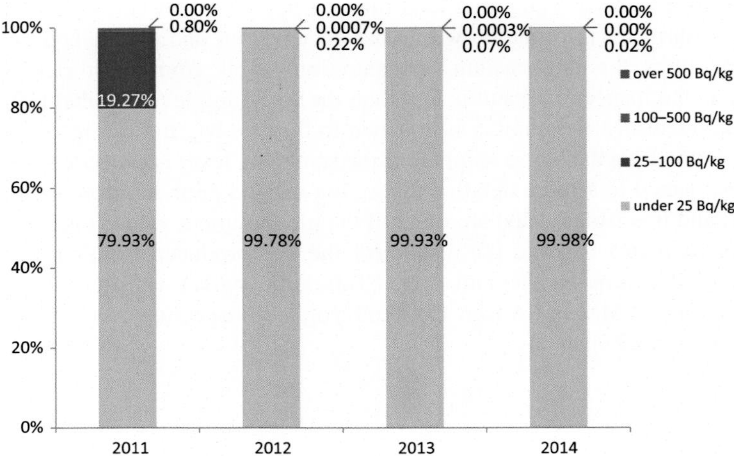

Fig. 3.3 Ratio of radiocesium concentration in rice for 4 years

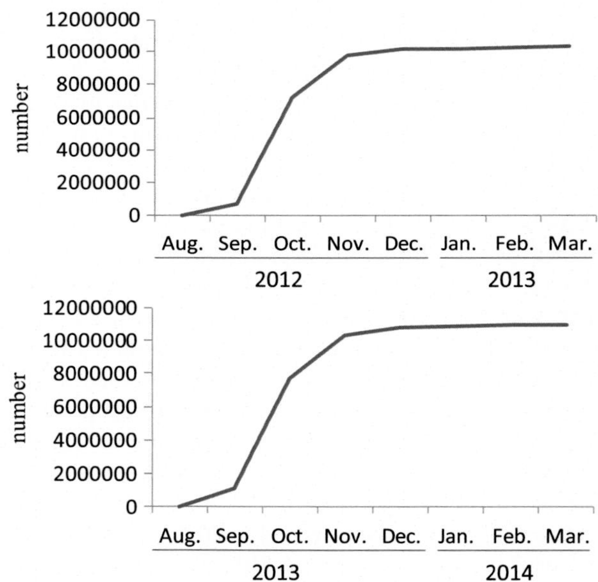

Fig. 3.4 Number of the inspection of all rice

period in a given day, two bags were inspected every minute. It is indicated that the belt conveyor tester can be considered an effective method for screening.

Rice is the main staple food of the Japanese diet, and it is the most valuable agricultural product in Fukushima Prefecture. Therefore, after the nuclear accident at Fukushima Dai-ichi NPP, inspections were performed thoroughly for rice than for other agricultural products. Note that the proportion of rice with radiocesium concentrations exceeding 100 Bq/kg was 0.8 % in 2011 and dropped to a mere 0.0007 % (71 bags out of the total 10,338,000), 0.0003 % (28 bags out of the total 11,001,000) and 0.00002 % (2 bags out of 10,988,824) in 2012, in 2013 and in 2014, respectively. In future, as agricultural operations restart in areas where planting is currently restricted, securing the safety of rice by thorough inspections and accurately communicating the inspection results will continue to be critical tools to reassure the public about food safety.

References

Chino M, Nakayama H, Nagai H, Terada H, Katata G, Yamazawa H (2011) Preliminary estimation of release amounts of 131I and 137Cs accidentally discharged from the Fukushima Daiichi nuclear power plant into the atmosphere. J Nucl Sci Technol 48:1129

Hirose K (2012) 2011 Fukushima Dai-ichi nuclear power plant accident: summary of regional radioactive deposition monitoring results. J Environ Radioact 111:13

Kinoshita N, Sueki K, Sasa K, Kitagawa J, Ikarashi S, Nishimura T, Wong YS, Satou Y, Handa K, Takahashi T, Sato M, Yamagata T (2011) Assessment of individual radionuclide distributions from the Fukushima nuclear accident covering Central-East Japan. Proc Natl Acad Sci USA 108:19526

Tensho K, Yeh KL, Mitsui S (1961) The uptake of strontium and cesium by plants from soil with special reference to the unusual cesium uptake by lowland rice and its mechanism. Soil Plant Food 6:176

Tsumura A, Komamura M, Kobayashi H (1984) Behavior of radioactive Sr and Cs in soils and soil-plant systems. Bull Natl Inst Agric Sci Ser B 36:57 (in Japanese)

Yasunari TJ, Stohl A, Hayano RS, Burkhart JF, Eckhardt S, Yasunari T (2011) Cesium-137 deposition and contamination of Japanese soils due to the Fukushima nuclear accident. Proc Natl Acad Sci USA 108:19530

Zheng J, Tagami K, Bu W, Uchida S, Watanabe Y, Kubota Y, Fuma S, Ihara S (2014) 135Cs/137Cs isotopic ratio as a new tracer of radiocesium released from the Fukushima nuclear accident. Environ Sci Technol 48:5433

Chapter 4
Cesium Accumulation in Paddy Field Rice Grown in Fukushima from 2011 to 2013: Cultivars and Fertilization

Yoshihiro Ohmori, Nobuhiro Tanaka, and Toru Fujiwara

Abstract After the accident at the Fukushima Daiichi Nuclear Power Plant, radioactive cesium (Cs) was released and the agricultural fields in Fukushima were contaminated. It became important to obtain data for radioactive Cs accumulation in rice grown in contaminated fields. We conducted a 3-year investigation in a Fukushima paddy field of radioactive Cs concentrations in various rice cultivars, and in two commercial rice cultivars grown under four different nutrient conditions. Our studies demonstrated substantial variation in radioactive Cs concentrations among the rice cultivars, and an increase in radioactive Cs concentrations in straw and brown rice under high nitrogen and low potassium conditions. Our 3-year investigations of radioactive Cs-contaminated rice in Fukushima paddy field shows that the rice grown in Fukushima is now well-monitored and contains much less than the allowed levels of radiation (100 Bq kg^{-1}).

Keywords Radioactive cesium • Straw • Brown rice • Rice cultivars • Fertilizer effect • Fukushima paddy field

4.1 Introduction

The accident at the Fukushima Dai-ichi Nuclear Power Plant in March 2011 released radionuclides to the broader area including the paddy fields around the nuclear power plant. The radioactive isotopes of cesium (Cs) have relatively long half-lives among the released radionuclides (2.06 years for ^{134}Cs and 30.2 years for ^{137}Cs) (Matsumura et al. 2011). Contamination of agricultural products by radioactive Cs will thus be a serious problem for a long time. We have previously reported the accumulation of radioactive Cs among different rice cultivars, and the

Y. Ohmori • N. Tanaka • T. Fujiwara (✉)
Laboratory of Plant Nutrition and Fertilizers, Department of Applied Biological Chemistry, Graduate School of Agricultural and Life Sciences, The University of Tokyo, Yayoi, Bunkyo-ku, Tokyo 113-8657, Japan
e-mail: atorufu@mail.ecc.u-tokyo.ac.jp

T.M. Nakanishi, K. Tanoi (eds.), *Agricultural Implications of the Fukushima Nuclear Accident*, DOI 10.1007/978-4-431-55828-6_4

effects of fertilizer on the accumulation of radioactive Cs in rice (Ohmori et al. 2014a, b). Here, we review these reports.

Cesium is an alkali metal, which is absorbed from the soil by roots and transported to various parts of rice plants, such as the brown rice and straw, which are served as foods for human and livestock, respectively. To reduce the Cs accumulation in rice, we need to understand the mechanism of Cs uptake and transportation in rice plants. Cs accumulation in rice is thought to be determined by both genetic and environmental factors. We measured the radioactive Cs concentration in 85 rice cultivars to find the genetic factors, and we investigated the effects of fertilizer on radioactive Cs accumulation in rice plants to reveal the environmental factors.

4.2 Difference in Radioactive Cesium Accumulation Among Rice Cultivars Grown in the Paddy Field at Fukushima from 2011 to 2013

4.2.1 Radioactive Cesium Accumulation Among 85 Rice Cultivars Grown in Fukushima Paddy Fields in 2011

Cesium is an alkali metal that is not essential to plant growth, but is toxic. Potassium (K) also belongs to alkali metal group, and it is an essential element for plant growth. It is believed that both Cs uptake and transport are mediated by K transporters. In *Arabidopsis*, one of the KUP/HAK/KT type transporters, AtHAK5, plays a role in non-radioactive cesium (^{133}Cs) absorption under low K conditions (Qi et al. 2008). In addition, *AtCNGC1* is a candidate gene for determining the natural variation of Cs concentrations (Kanter et al. 2010). However, the whole mechanism of Cs uptake and transport remains unclear.

To reduce Cs accumulation in rice, it is crucial to understand that there is variation in Cs uptake among different rice cultivars. The difference in ^{133}Cs concentrations in brown rice among different rice cultivars has been reported (Yamaguchi et al. 2012), and the concentration of ^{137}Cs in rice grown in Aomori Prefecture before the Fukushima accident has also been reported (Tsukada et al. 2002). However, the amount of radioactive Cs fallout from the Fukushima Dai-ichi Nuclear Power Plant after the earthquake in 2011 was much higher than that derived from past fallout. Thus, a reinvestigation of the accumulation levels of radioactive Cs in rice is needed in the Fukushima area.

We selected 85 rice cultivars from the World Rice Core Collection (WRC), the Japanese Rice Landrace Mini Core Collection (JRC), and other domestic varieties (Table 4.1). The WRC consists of 67 varieties and covers 91 % of the genetic variation in about 37,000 rice landraces. The JRC consists of 50 varieties and covers 87.5 % of genetic variation in about 2000 Japanese rice landraces (Ebana et al. 2008; Kojima et al. 2005). We planted the 85 rice cultivars in the Fukushima paddy field on May 31st, 2011, and harvested them on September 23rd, October

Table 4.1 List of the 85 rice cultivars tested in this study

Number	Cultivar name	Number	Cultivar name
1	Karahoushi	44	Kabashiko
2	Houmanshindenine	45	Jamaica
3	Mansaku	46	Shichimenchou Mochi
4	Himenomochi	47	Khauk Yoe
5	Akage	48	Tachisugata
6	Hassokuho	49	Shiroine
7	Kahei	50	Mizuhochikara
8	Shinyamadaho 2	51	Akamai
9	Aichiasahi	52	Kusanohoshi
10	Hamasari	53	Sekiyama
11	Hakamuri	54	Fukoku
12	Shinriki Mochi	55	Shinriki
13	Raiden	56	Leaf Star
14	Puluik Arang	57	Chinya
15	Ginbouzu	58	Gaisen Mochi
16	Vary Futsi	59	Meguro Mochi
17	Ishijiro	60	Senshou
18	Nipponbare	61	Moritawase
19	Mogumoguaoba	62	Momiroman
20	Bekogonomi	63	Yamada Baka
21	Nishiaoba	64	Hetadawee
22	Nagoyashiro	65	Taichung 65
23	Aikoku	66	Mack Kheua
24	Kameji	67	Rikutou Rikuu 2
25	Yumeaoba	68	Omachi
26	Moroberekan	69	Chinsurah Boro 2
27	Hosogara	70	Hiyadachitou
28	Kasalath	71	Okka Mososhi
29	Kyoutoasahi	72	Daw Dam
30	Co 13	73	Deng Pao Zhai
31	Tachiaoba	74	Basilanon
32	Bekoaoba	75	Koshihikari
33	Kusahonami	76	Badari Dhan
34	Dango	77	Hoshiaoba
35	Tupa 121-3	78	Asominori
36	Hirayama	79	Kaneko
37	Naba	80	Oiran
38	Hinode	81	Khau Mac Kho
39	Muha	82	Joushuu
40	Touboshi	83	Wataribune
41	Bouzu Mochi	84	Iruma Nishiki
42	Fukuhibiki	85	Shinshuu
43	Okabo		

Table 4.2 Mean (standard deviation), median, and range for cesium concentration in different rice cultivars grown at Fukushima in 2011

Cs concentration in straw					Cs concentration in brown rice				
	Mean	(SD)	Median	Range		Mean	(SD)	Median	Range
^{134}Cs (Bq/kg)	38.9	(13.5)	35.8	19.4 ∽ 73.4	^{134}Cs (Bq/kg)	8.1	(4.7)	6.7	0.7 ∽ 20.3
^{137}Cs (Bq/kg)	39.0	(17.2)	35.5	10.3 ∽ 100.3	^{137}Cs (Bq/kg)	11.6	(5.8)	10.2	2.7 ∽ 26.6

4th, and October 18th. The radioactive Cs concentrations in harvested rice straw and brown rice were independently determined. The ^{134}Cs and ^{137}Cs concentrations in the straw were 19.4–73.4 and 10.3–100.3 Bq kg^{-1}, respectively (Table 4.2). In addition, the mean concentrations were 38.9 and 39.0 Bq kg^{-1}, respectively, and the medians were 35.8 and 35.5 Bq kg^{-1}, respectively (Table 4.2). In brown rice, the ^{134}Cs and ^{137}Cs concentrations were 0.7–20.3 Bq kg^{-1} and 2.7–26.6 Bq kg^{-1}, respectively (Table 4.2). The means were 8.1 and 11.6 Bq kg^{-1}, respectively, and the medians were 6.7 and 10.2 Bq kg^{-1}, respectively (Table 4.2). Both the straw and brown rice from the selected rice cultivars showed a large variation in radioactive Cs concentration. This variation can be used to isolate either Cs uptake or transport-related factors.

Next, we correlated the Cs concentration between straw and brown rice among the 85 rice cultivars. Both ^{134}Cs and ^{137}Cs concentrations correlated significantly and positively between straw and brown rice (Fig. 4.1), that are $p = 1.2 \times 10^{-6}$ and

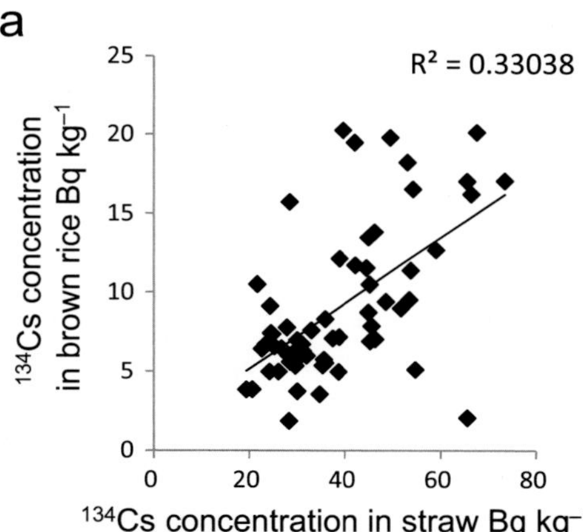

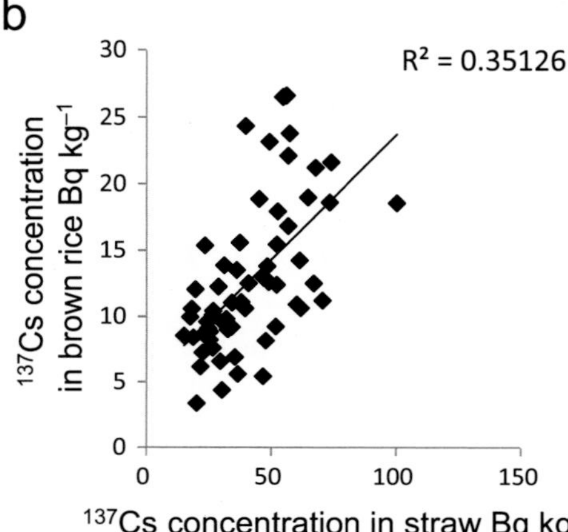

Fig. 4.1 Correlation diagram for Cs concentration. (**a**) Straw versus brown rice for ^{134}Cs. (**b**) Straw versus brown rice for ^{137}Cs. The corresponding coefficients of determination (R^2) are shown. The *black line* represents the linear regression line corresponding to the least square adjustment of all the data

$p = 4.9 \times 10^{-7}$ for ^{134}Cs and ^{137}Cs, respectively (Fig. 4.1). The coefficients of determination (R^2) were 0.33 and 0.35 for ^{134}Cs and ^{137}Cs, respectively. Thus, we concluded that the Cs concentrations in brown rice might be estimated from the Cs concentrations in straw, although there were some exceptions.

4.2.2 Radioactive Cesium Accumulation Among 15 Selected Rice Cultivars Grown in a Fukushima Paddy Field in 2012 and 2013

On the basis of the Cs concentration of brown rice in 2011, we selected 15 rice cultivars to test the reproducibility of Cs uptake, and planted them at a Fukushima paddy field in 2012 and 2013. Khau Mac Kho, Asominori, Kaneko, and Deng Pao Zhai were selected as high Cs accumulating cultivars; whereas, Kasalath, Hamasari, Kameji, Aichiasahi, Wataribune, Mansaku, Akage, and Hassokuho were selected as low Cs accumulating cultivars. In addition, Koshihikari, Nipponbare, and Taichung 65 were selected as typical Japanese cultivars. In 2012, we planted the selected 15 rice cultivars at a Fukushima paddy field on May 23rd, and sampled them on October 13. In 2013, the planting and harvesting dates were May 14th and October 10th, respectively. Khau Mac Kho, Asominori, and Deng Pao Zhai showed relatively higher concentrations of ^{137}Cs in brown rice among different rice cultivars (Fig. 4.2). On the other hand, Hamasari, Aichiasahi, and Mansaku showed relatively lower ^{137}Cs concentrations in brown rice compared with the other cultivars (Fig. 4.2). These results were comparatively conserved in 3-year investigations.

Our results significantly provide data for Cs accumulation levels among different rice cultivars in a Fukushima paddy field. A molecular genetic approach to rice cultivars with different Cs accumulation may enable identification of genes that regulate Cs uptake and transportation in rice.

4.3 Fertilizer Effects on Cs Accumulation in Rice

4.3.1 General Information of Fertilizer Effects on Cs Accumulation in Plants

Both K and Cs are alkali metals, and Cs transportation is known to be mediated by several K transporters (Qi et al. 2008; Jabnoune et al. 2009). Thus, Cs uptake and transportation by K transporters compete with K uptake and transportation in plants. It has been reported that Cs uptake is enhanced under low K conditions in various plant species (Shaw 1993). K fertilizer applications can reduce Cs absorption in crops such as wheat, barley, rye, and potato under K deficient conditions

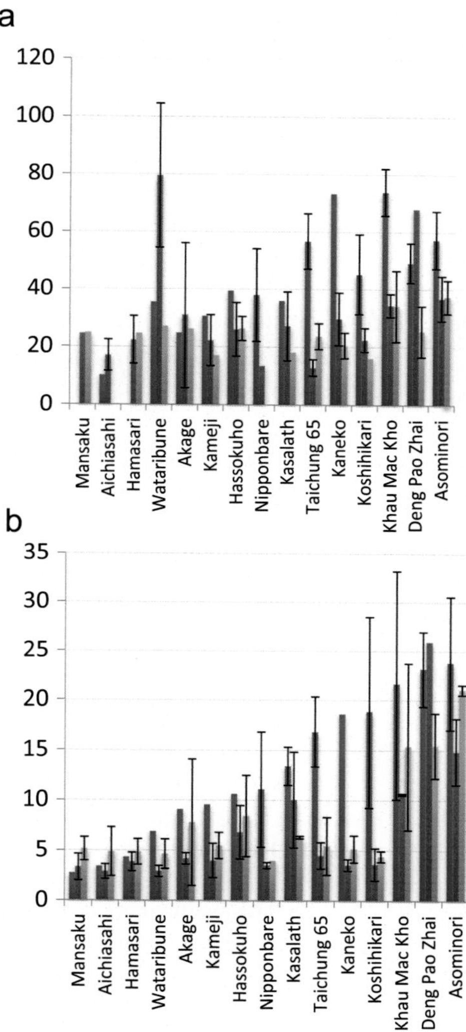

Fig. 4.2 Comparison of ^{137}Cs concentrations among 2011–2013 data in straw and brown rice. (**a**) ^{137}Cs concentrations (Bq kg^{-1}) in straw from selected cultivars. (**b**) ^{137}Cs concentrations (Bq kg^{-1}) in brown rice from selected cultivars. *Blue, red,* and *green boxes* indicate 2011, 2012, and 2013 data, respectively. Means and standard deviations are shown (n = 3)

(Lemmbrechts 1993). On the other hand, fertilizer application has little effect on Cs absorption under adequate K conditions.

Ammonium (NH_4^+) is known to affect the elution of Cs from soil by replacing NH_4^+ with Cs^+. Therefore, high concentrations of NH_4^+ in the soil enhance Cs^+ elution, resulting in the promotion of Cs^+ absorption by plants. It has been reported that the application of nitrogen fertilizer enhances Cs uptake by plants in the field, although the degree of this effect depends on the soil type and other conditions (Lemmbrechts 1993; Smolders et al. 1997).

Before March 11, 2011, studies on radioactive Cs accumulation had been mainly conducted at the site of the Chernobyl Nuclear Power Plant accident in Russia.

Therefore, the behavior of radioactive Cs affected by fertilizers in paddy fields and andosols, which are the common field condition and soil-type in Japan, were not well elucidated. Thus, it is important to investigate the effects of K and N fertilizer on radioactive Cs absorption in rice grown in Japanese paddy fields.

In the next chapter, we will describe radioactive Cs concentrations in rice grown in paddy fields under four different fertilizer conditions at Ishidairayama, Yamakiya, Kawamata-cho in Fukushima in 2011 and 2012.

4.3.2 Radioactive Cs Concentrations in Rice Grown in Paddy Fields Under Four Different Fertilizer Conditions at Fukushima

To investigate the effect of fertilizer applications on radioactive Cs concentrations in rice, we cultured two commercial rice cultivars, Koshihikari and Hitomebore, in 2011 and 2012 under four different fertilizer conditions: normal, −K, −K + 2N, and no fertilizer. Under normal condition, a commercial fertilizer containing 8:18:16 (N:P:K; equivalent to 6, 9, and 8 kg per 10 a) was applied as a basal fertilizer. Under K-depleted conditions, N and P were given as urea and monocalcium phosphate, respectively. Under −K condition, N, P, and K were given as 6, 9, and 0 kg per 10 a, respectively. Under −K + 2N condition, N, P, and K were given as 12, 9 and 0 kg per 10 a, respectively. Under no fertilizer condition, no fertilizers were applied.

4.3.2.1 Radioactive Cs Concentration in Rice Straw Grown in a Paddy Field at Kawamata-cho

To assess the effect of fertilizer conditions on radioactive Cs concentrations in rice, we determined the radioactive Cs (^{134}Cs and/or ^{137}Cs) concentrations in straw harvested at the ripening stage.

In 2011, the ^{134}Cs concentration in straw under the −K + 2N condition was 1.5 times higher than that under the normal condition (Fig. 4.3a). The ^{134}Cs concentrations in straw were also high under the −K condition compared to those under the normal condition (Fig. 4.3a). On the other hand, there was no difference in the ^{134}Cs concentrations in straw under the normal and no fertilizer conditions. Similar trends were also observed for the ^{137}Cs concentrations (Fig. 4.3b).

In 2012, we replanted Koshihikari and Hitomebore at the same paddy field and investigated the reproducibility of the ^{137}Cs concentrations in the straw. The ^{137}Cs concentrations in the straw were highest under the −K + 2N condition (Fig. 4.3c). The ^{137}Cs concentrations in straw were also high under the −K condition (Fig. 4.3c). Under the no fertilizer condition, the ^{137}Cs concentrations in straw were similar to those under the normal fertilizer condition. All patterns of

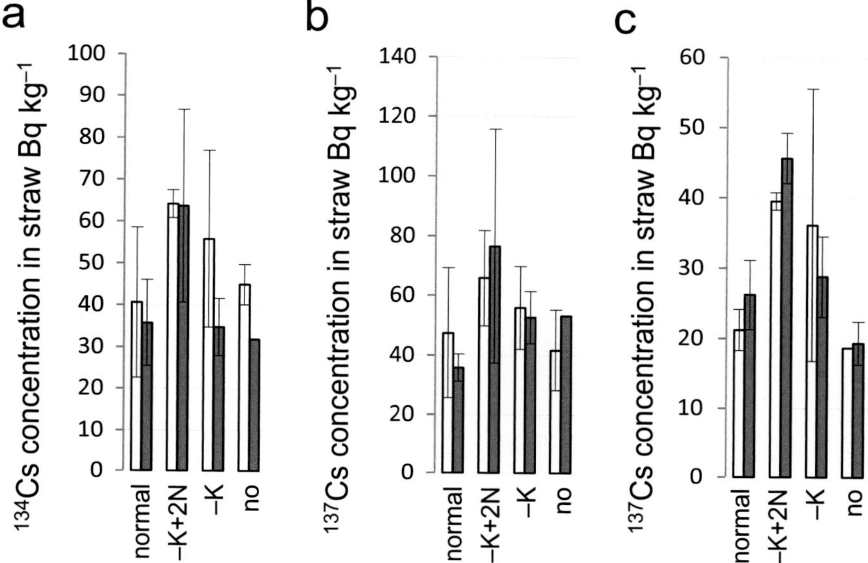

Fig. 4.3 Cesium concentrations in straw under different fertilizer conditions in 2011 and 2012. (**a**) ^{134}Cs concentrations in 2011. (**b**) ^{137}Cs concentrations in 2011. (**c**) ^{137}Cs concentrations in 2012. *White and black boxes* indicate Koshihikari and Hitomebore, respectively. Means and standard deviations are shown (n = 3). The concentrations are presented on a dry-weight basis. Normal, the normal fertilizer condition; –K + 2 N, twofold nitrogen without potassium condition; –K, normal nitrogen and phosphorus but no potassium condition; and no, no fertilizer condition

radioactive Cs concentrations in rice straw were similar between Koshihikari and Hitomebore in 2011 and 2012 (Fig. 4.3).

4.3.2.2 Radioactive Cs Concentrations in Brown Rice Grown in a Paddy Field at Kawamata-cho

We determined the radioactive Cs concentrations in brown rice grown at Kawamata-cho in 2011 and 2012. In 2011, the trends for radioactive Cs accumulation in brown rice were very similar between ^{134}Cs and ^{137}Cs (Fig. 4.4a, b). The radioactive Cs concentrations in brown rice were highest under the −K + 2N condition, being about twice that under the normal condition. Under the −K condition, the radioactive Cs concentrations in brown rice were also higher than those under the normal condition in Koshihikari (Fig. 4.4a, b). The radioactive Cs concentrations in brown rice were the lowest under the no fertilizer condition.

In 2012, the trends of ^{137}Cs accumulation in brown rice under the four different fertilizer conditions were very similar to those observed in 2011 (Fig. 4.4). The ^{137}Cs concentrations in brown rice were highest under the −K + 2N condition, and lowest under the no fertilizer condition both in 2011 and 2012 (Fig. 4.4b, c).

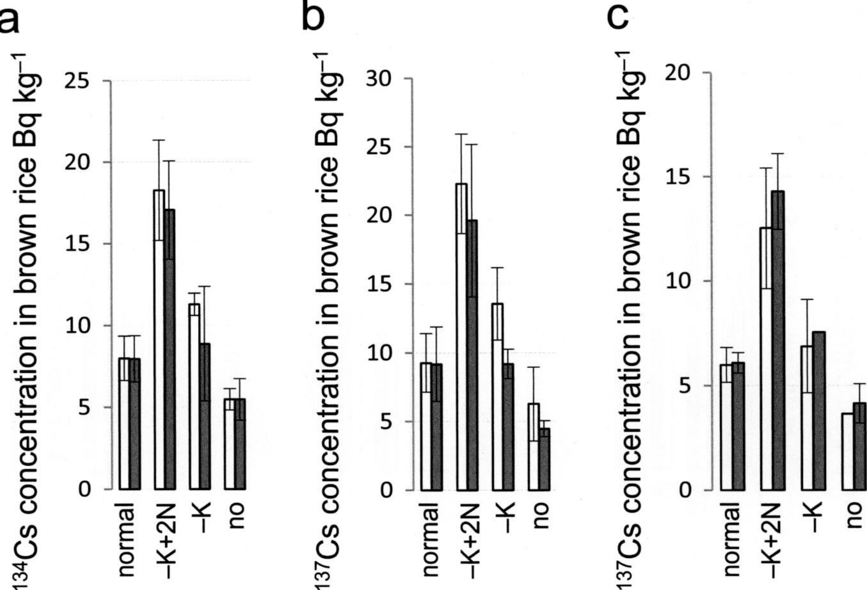

Fig. 4.4 Cesium concentrations in brown rice under different fertilizer conditions in 2011 and 2012. (**a**) ^{134}Cs concentrations in 2011. (**b**) ^{137}Cs concentrations in 2011. (**c**) ^{137}Cs concentrations in 2012. *White and black boxes* indicate Koshihikari and Hitomebore, respectively. Means and standard deviations are shown (n = 3). Fertilizer conditions are the same as described in the legend of Fig. 4.3

However, in contrast to the results of 2011, the ^{137}Cs accumulation in brown rice under the −K condition in 2012 was no different to that under the normal condition (Fig. 4.4c).

In our study, low K conditions tended to increase the radioactive Cs concentrations both in straw and brown rice grown in the contaminated paddy field at Kawamata-cho in Fukushima. This result may be caused by chemical competition between K and Cs. In addition, it is noteworthy that nitrogen fertilizer affects radioactive Cs concentrations in rice. The fertilizer condition that caused the highest radioactive Cs concentration in rice was −K + 2N in both 2011 and 2012. This result suggests that not only K fertilizer, but also N fertilizer affects radioactive Cs concentrations in rice grown in a Japanese paddy field.

The mechanism that increases radioactive Cs concentrations in rice by N fertilizer application is still unknown. One hypothesis is that N fertilizers elute radioactive Cs from the soil surface and enhance the radioactive Cs uptake by rice. To avoid unexpectedly high-levels of radioactive Cs in rice (over 100 Bq kg^{-1}; the governmental new safety standards for radioactive Cs in food products in Japan), we may need to implement the N and K fertilizer conditions.

References

Ebana K, Kojima Y, Fukuoka S, Nagamine T, Kawase M (2008) Development of mini core collection of Japanese rice landrace. Breed Sci 58:281–291

Jabnoune M, Espeout S, Mieulet D, Fizames C, Verdeil JL, Conéjéro G, Rodríguez-Navarro A, Sentenac H, Guiderdoni E, Abdelly C, Véry AA (2009) Diversity in expression patterns and functional properties in the rice HKT transporter family. Plant Physiol 150:1955–1971

Kanter U, Hauser A, Michalke B, Dräxl S, Schäffner AR (2010) Caesium and strontium accumulation in shoots of *Arabidopsis thaliana*: genetic and physiological aspects. J Exp Bot 61:3995–4009

Kojima Y, Ebana K, Fukuoka S, Nagamine T, Kawase M (2005) Development of an RFLP-based rice diversity research set of germplasm. Breed Sci 55:431–440

Lemmbrechts J (1993) A review of literature on the effectiveness of chemical amendments in reducing the soil-to-plant transfer of radiostrontium and radiocaesium. Sci Total Environ 137:81–98

Matsumura H, Saito K, Ishioka J, Uwamino Y (2011) Diffusion of radioactive materials from Fukushima-Daiichi nuclear power station obtained by gamma-ray measurements on expressways. Trans Atomic Energy Soc Jpn 10:152–162 (in Japanese)

Ohmori Y, Inui Y, Kajikawa M, Nakata A, Sotta N, Kasai K, Uraguchi S, Tanaka N, Nishida S, Hasegawa T, Sakamoto T, Kawara Y, Aizawa K, Fujita H, Li K, Sawaki N, Oda K, Futagoishi R, Tsusaka T, Takahashi S, Takano J, Wakuta S, Yoshinari A, Uehara M, Takada S, Nagano H, Miwa K, Aibara I, Ojima T, Ebana K, Ishikawa S, Sueyoshi K, Hasegawa H, Mimura T, Mimura M, Kobayashi NI, Furukawa J, Kobayashi D, Okouchi T, Tanoi K, Fujiwara T (2014a) Difference in cesium accumulation among rice cultivars grown in the paddy field in Fukushima prefecture in 2011 and 2012. J Plant Res 127:57–66

Ohmori Y, Kajikawa M, Nishida S, Tanaka N, Kobayashi NI, Tanoi K, Furukawa J, Fujiwara T (2014b) The effect of fertilization on cesium concentration of rice grown in a paddy field in Fukushima prefecture in 2011 and 2012. J Plant Res 127:67–71

Qi Z, Hampton CR, Shin R, Barkla BJ, White PJ, Schachtman DP (2008) The high affinity K^+ transporter AtHAK5 plays a physiological role in planta at very low K^+ concentrations and provides a caesium uptake pathway in *Arabidopsis*. J Exp Bot 59:595–607

Shaw G (1993) Blockade by fertilisers of caesium and strontium uptake into crops: effects on the root uptake process. Sci Total Environ 137:119–133

Smolders E, Van den Brande K, Merckx R (1997) Concentrations of ^{137}Cs and K in soil solution predict the plant availability of ^{137}Cs in soils. Environ Sci Technol 31:3432–3438

Tsukada H, Hasegawa H, Hisamatsua S, Yamasakib S (2002) Transfer of ^{137}Cs and stable Cs from paddy soil to polished rice in Aomori, Japan. Environ Pollut 59:351–363

Yamaguchi N, Takata Y, Hayashi K, Ishikawa S, Kuramata M, Eguchi S, Yoshikawa S, Sakaguchi A, Asada K, Wagai R, Makino T, Akahane I, Hiradate S (2012) Behavior of radio cesium in soil-plant system and its controlling factor: a review. Rep Natl Inst Agro Environ Sci Jpn 31:75–129 (in Japanese)

Chapter 5
Physiological Verification of the Effect of Potassium Supply on the Reduction of Radiocesium Content in Rice Grain

Natsuko I. Kobayashi

Abstract To achieve the reduction of the radiocesium content in rice grain, the application of potassium (K) fertilizer to the paddy field is currently recommended. However, physiological basis of the effect of the K addition on the radiocesium accumulation in rice has not been enough provided. Here, the uptake and the distribution of cesium (Cs) in rice plant grown in either K-deficient or K-sufficient hydroponic medium containing ^{137}Cs are presented aiming to describe the exact impact of K fertilization on the Cs behavior within the plant. In the K-deficient plant, the amount of ^{137}Cs accumulated in the rice grain was 10 times as much as the K-sufficient rice. The determination of ^{137}Cs content as well as other cationic ions in each part of the rice showed the intensive transport of ^{137}Cs to the ear part composed of brown rice, husk and culm, in which K was also accumulated. It could supposed that Cs transport is regulated basically similarly to the K transport within the plant body. Then, K fertilization is suggested to reduce the Cs content in rice grain efficiently through the reduction of Cs uptake in the roots and Cs accumulation to the ear part.

Keywords Oryza sativa • Paddy field • Potassium fertilization • Tracer • Uptake • Transport

5.1 Introduction

Among agricultural products, rice contaminated with radiocesium has received particular attention because it is the staple food of Japan and the main agriproduct in Fukushima. After 2011, an extensive survey on rice was conducted and the result showed a clear inverse relationship between soil potassium (K) and radiocesium concentrations (http://www.maff.go.jp/j/kanbo/joho/saigai/pdf/kome.pdf). In addition, when the K fertilizer was applied to paddy fields where rice containing

N.I. Kobayashi (✉)
Graduate School of Agricultural and Life Sciences, The University of Tokyo, 1-1-1 Yayoi, Bunkyo-ku, Tokyo 113-8657, Japan
e-mail: anikoba@mail.ecc.u-tokyo.ac.jp

T.M. Nakanishi, K. Tanoi (eds.), *Agricultural Implications of the Fukushima Nuclear Accident*, DOI 10.1007/978-4-431-55828-6_5

relatively high amounts of radiocesium was produced, the brown rice produced the following year had much lower radiocesium concentrations. Actually, many previous physiological experiments have indicated that high K concentration in the rhizosphere could reduce Cs absorption by the roots in several plant species (Zhi et al. 2002; Robison et al. 2009). This reduction was thought to be due to the similar chemical properties of K and Cs. Because they are both alkali metals, Cs is assumed to enter the root cells through the K transport apparatus, which is where the competition between these elements is thought to occur.

The competition between K uptake and Cs uptake in rice roots is recently described (Kobayashi et al. 2015). The kinetics of K uptake was directly analyzed using ^{42}K, and simultaneously, the uptake rate of Cs was calculated using ^{137}Cs. As the K concentration in the uptake medium increased, the K uptake rate increased and the Cs uptake rate decreased (Kobayashi et al. 2015). In *Arabidopsis* plants, the molecule mediating K and Cs uptake has been identified (Qi et al. 2008).

In addition to the relationship between soil K concentration and radiocesium contamination of rice, an intriguing observation about radiocesium distribution in rice plants was found in 2011. In some paddy fields in Fukushima, brown rice containing over 500 Bq/kg of radiocesium was produced. We analyzed radiocesium distribution in rice seedlings harvested in those paddy fields (Paddy-field A) and found that younger organs, such as the ear and the first and second internodes, accumulated more radiocesium (Fig. 5.1). The leaf with the highest radiocesium concentration was the uppermost leaf (Fig. 5.1). In contrast, the older leaves contained larger amounts of radiocesium for rice plants harvested in the paddy field (Paddy field B) where brown rice without radiocesium contamination (<4 Bq/kg) was produced (Fig. 5.1). Given that the soil K concentration in Paddy-field A was low, radiocesium accumulation in the younger parts could be considered to be triggered by K deficiency. There is frequent K movement between plant parts. When the plant encounters K shortage, K can be translocated from the older tissues to the younger tissues to maintain growth. These K movements inside the plant could be assumed to be mimicked by Cs; thus, Cs as well as K might accumulate in younger tissues in response to K deficiency. To produce rice with the least radiocesium content in paddy fields, it is important to understand the radiocesium distribution in rice plants and to distinguish the influencing factors. Therefore, apart from environmental factors, we investigated the physiological effect of K supply on Cs uptake and transport in rice plants using ^{137}Cs.

5.2 Effect of K Concentration in Nutrient Solution on Cs Distribution in Rice Plants

To analyze the effect of K supply on ^{137}Cs behavior, we compared ^{137}Cs distribution in rice plants grown with or without K. Rice seedlings (*Oryza sativa* L. Nipponbare) were grown in half-strength Kimura B nutrient solution for

Fig. 5.1 Radiocesium distribution in rice plants harvested in Fukushima in 2011. The *upper picture* shows the rice plant harvested in paddy-field A where the highly contaminated rice grain (approximately 500 Bq/kg) was produced. The rice plant was separated into organs and placed with clods of paddy-soil and some reference samples (surrounded with frames). Radioactivity was detected using an imaging plate (BAS IP MS, FujiFilm) and was described with a false color. *Arrows* indicate the internodes. The *bottom graph* shows the concentration of ^{137}Cs in the leaves of the rice plants harvested from Paddy-field A and B [Modified from the report by Tanoi et al. (2013)]

3 weeks and then transplanted either to 3 mM K or K-free nutrient solution containing ^{137}Cs (9 kBq/L). The K-sufficient and K-deficient rice plants were grown for another 8 weeks until maturity. For cultivation, a plant growth chamber was set at 30 °C with a daily 12 h light and 12 h dark cycle. After harvesting, the rice plants were separated into several parts and the radioactivity of ^{137}Cs in each part was measured to determine the ^{137}Cs distribution. The K-deficient rice contained nearly 3-times the amount of ^{137}Cs compared with the K-sufficient rice. As presented in Fig. 5.2, it was clear that the K-sufficient rice accumulated ^{137}Cs in the older leaves, whereas the K-deficient rice contained a large amount of ^{137}Cs in the ear and the culm, followed by the upper leaves. Therefore, the distribution of ^{137}Cs in the K-deficient rice (Fig. 5.2) was similar to the radiocesium distribution found in the rice grown in Paddy-field A (Fig. 5.1). In K-deficient rice, the ^{137}Cs accumulated in the ear accounted for more than 25 % of ^{137}Cs found in the shoots, whereas it was less than 10 % in the K-sufficient rice (Fig. 5.2). As a result, K concentration in the culture solution was shown to impact significantly on Cs distribution within a rice plant.

5.3 Cation Concentration in K-Sufficient and K-Deficient Rice Plants

Does K concentration in solution alter the Cs distribution specifically? This question is important for considering the mechanism regulating Cs transport inside rice plants. Therefore, we investigated the distribution of sodium (Na), magnesium (Mg), calcium (Ca), as well as K and ^{137}Cs in K-sufficient and K-deficient rice plants (Fig. 5.3). The concentration of K in the leaves was higher than in the brown rice when K was sufficient, which was the reversed response to K-starvation. Interestingly, this alteration was also observed for ^{137}Cs concentrations. K concentration in the sink organs, such as brown rice, husk, and culm, was not altered by K deficiency and the order of concentration was brown rice < husk < culm in both K-sufficient and K-deficient rice plants (Fig. 5.3). K was found to be actively transported from the leaves to the reproductive organs to maintain their K concentration, even if K was not supplied. The concentration of ^{137}Cs was in the order of brown rice < husk < culm in both K-sufficient and K-deficient rice plants, which was very similar to the order of K concentration, and the ^{137}Cs concentration in these reproductive organs co-increased in response to K shortage (Fig. 5.3). Less drastically, K-starvation was shown to cause an increase in Mg concentration and a decrease in Ca concentration, although the distribution of Mg and Ca among organs was not largely modified (Fig. 5.3). On the other hand, Na accumulation in the leaves was promoted drastically under K deficiency. Previous reports suggested that the additional Na accumulated in the K deficient leaves could compensate for some function of K, and this could be one reason why Na absorption was activated under K deficiency (Rodriguez-Navarro 2000). However, unlike K and ^{137}Cs, Na

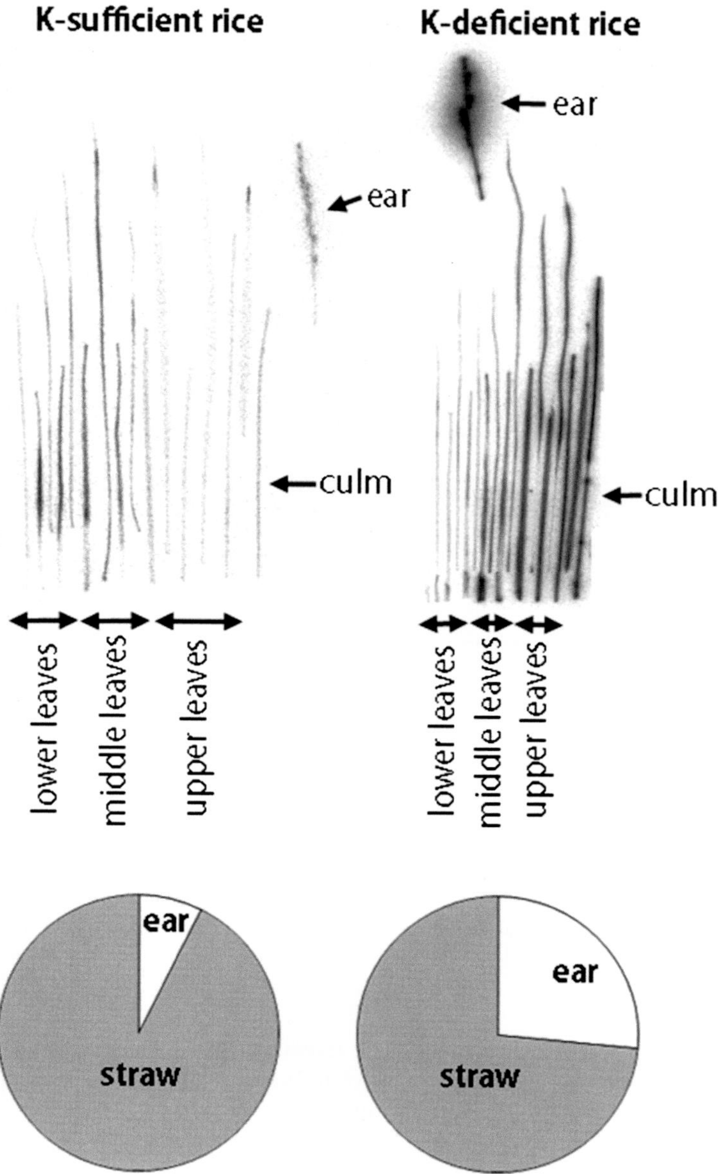

Fig. 5.2 The distribution of ^{137}Cs radioactivity among tissues at harvest. Rice seedlings hydroponically grown with K (3 mM, K-sufficient) or without K (0 mM, K-deficient) were harvested and separated into tissues to determine the ^{137}Cs content using an imaging plate. ^{137}Cs of 9 kBq/L was supplied from 3 weeks after germination until harvest. In the *top images*, ^{137}Cs radioactivity in each tissue was presented in the *gray-scale*. In K-deficient rice, *upper leaves* accumulated larger amounts of ^{137}Cs compared to the *lower leaves*. The *bottom graph* presents the distribution of ^{137}Cs between the ear and the straw [Modified from Kobayashi and Nobori (2014)]

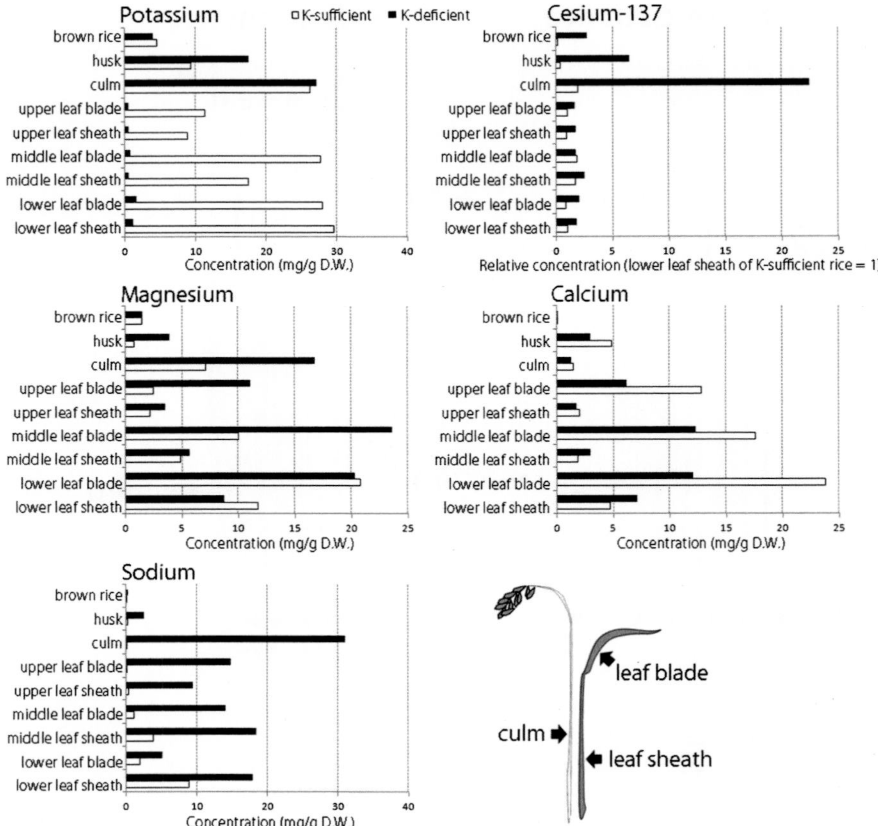

Fig. 5.3 Concentration of K, ^{137}Cs, Mg, Ca, and Na in each tissue of K-sufficient (*white bar*) and K-deficient (*black bar*) rice plants at harvest. To measure K, Mg, Ca, and Na, the tissues were digested with nitric acid and analyzed using ICP-OES (Optima 7300, PerkinElmer). The concentration of ^{137}Cs was determined using gamma counting [Modified from Kobayashi and Nobori (2014)]

concentration in the brown rice and husk remained lower than that in the leaves. These observations indicate that the mechanism regulating Cs allocation in rice plants is closely related to the K transport mechanism.

5.4 The Period for Cs Uptake

Given the effect of K supply on the reduction of Cs uptake into the root and Cs transport in brown rice plants, the application of K fertilizer in paddy fields is suggested as an effective measure to reduce the radiocesium content in rice.

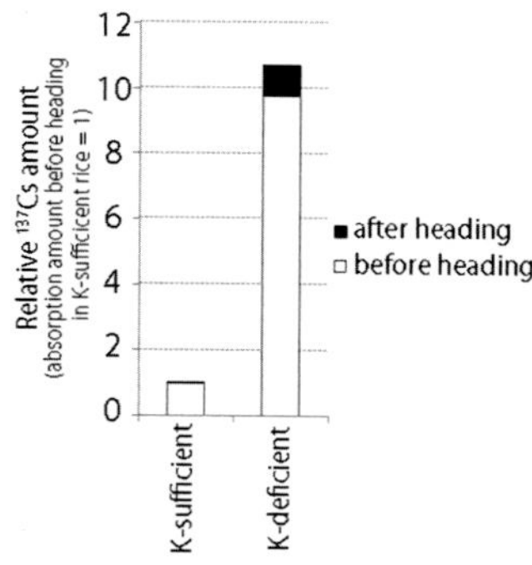

Fig. 5.4 The ^{137}Cs content in ears of K-sufficient and K-deficient rice at harvest and the period of ^{137}Cs absorption. K-deficient rice accumulated 10-times more ^{137}Cs in the ear as K-sufficient rice. Over 90 % of the ^{137}Cs in the ear at harvest, was absorbed before heading, and was probably relocated from the leaves to the ear as the ear matured, regardless of the K condition [Modified from Kobayashi and Nobori (2014)]

Subsequently, to develop a practical K fertilization scheme, it is important to determine the developmental period when rice plants absorb Cs and transport it toward the grain. Therefore, we compared the ^{137}Cs amount in the ear of rice plants supplied with ^{137}Cs after the third week until harvest with that of rice plants to which ^{137}Cs was supplied only after heading. The disparity between them could correspond to the ^{137}Cs amount absorbed before heading. The results showed that over 90 % of ^{137}Cs accumulated in the ear at harvest was absorbed before heading (Fig. 5.4). This means that most of the ^{137}Cs found in the ear was once stored in other parts, such as the leaves, and then relocated to the ear after heading. Furthermore, the amount of ^{137}Cs relocated to the ear after heading was increased nearly tenfold in response to K deficiency. Considering that the relocation of Cs from the leaves to the ear can accompany K movement, which could be modified depending on the level of K supply, it is possible that K supplement after heading can reduce Cs transport toward the ear. Thus, we tested this idea by transplanting K-deficient rice seedlings to the 3 mM K medium without ^{137}Cs at heading.

5.5 Evaluation of the Effect of Additional K Fertilization on Cs Movement

At harvest, the proportion of ^{137}Cs content in the ear to the total ^{137}Cs absorbed before heading (the ear fraction) in the K-sufficient rice and K-deficient rice was 17.3 % and 27.8 %, respectively (Table 5.1). Then, the ear fraction in rice plants to which K was supplied after heading was 25.0 % (Table 5.1). These results imply that K fertilization after heading has only a minor effect on Cs relocation. Meanwhile, transition to the K-rich condition after heading is assumed to be effective for reducing Cs uptake in roots. However, such a reduction is expected to have a limited impact on Cs content in brown rice because the amount of Cs absorbed

Table 5.1 Influence of K supply on ^{137}Cs transport to the ear after heading

K- condition		Distribution to the ear (%)
Before heading	After heading	
Sufficient	Sufficient	17.3
Deficient	Deficient	27.8
Deficient	Sufficient	25.0

^{137}Cs was added to the solution medium only before heading, and its distribution at harvest was determined. Rice plants grown under K sufficient conditions both before and after heading accumulated 17.3 % of ^{137}Cs in the ear. The percentage increased to 27.8 % in rice plants grown under K deficient conditions. Then, rice plants grown under K-deficient conditions until heading and then transplanted into K-sufficient solution relocated 25.0 % of ^{137}Cs to the ear [Modified from Kobayashi and Nobori (2014)]

after heading accounts for less than 10 % of the Cs content in the ear part at harvest (Fig. 5.4). Therefore, in consideration of plant physiology, the importance of a base-fertilizer, rather than additional fertilization, was demonstrated to reduce radiocesium contamination in rice.

5.6 Conclusion and Future Perspectives

The supply of K to rice plants decreased the Cs content in brown rice as a consequence of reduction of both Cs uptake by the root and Cs transport toward the edible part. Application of K fertilizer to some paddy fields in Fukushima apparently reduced the radiocesium concentration in brown rice, and the ear fraction of radiocesium content was decreased by half (unpublished data). Finally, in the autumn of 2014, all the brown rice commercially produced passed the test for sale. To keep the radiocesium contamination low, it could be important to maintain an appropriate K condition in rice plants. Regarding the determination of the K condition in plants, analyzing the radiocesium distribution among tissues, as well as the K concentration in the soil, is thought to improve the accuracy. If radiocesium concentration is lower in younger tissues than older tissues, the plant is considered to have enough K and thus additional K fertilization would have only minor benefits through the absorption competition as previously reported (Kobayashi et al. 2015). In these cases, some other factors influencing radiocesium behavior should be evaluated to reduce the radiocesium contamination.

In this study, we focused on the similarity of behavior between K^+ and Cs^+. However, their behavior is close but not the same. The ratio of K uptake rate to Cs uptake rate was found to be 7–11 times higher than the ratio of K concentration to Cs concentration in the culture medium, indicating that the root absorbs K selectively over Cs (Kobayashi et al. 2015). In addition, K was shown to be selectively relocated to the rice grain over Cs (Nobori et al. 2014). Characterization of the molecular machinery regulating the movement of K and Cs can further assist our understanding of Cs behavior in plants and assist breeding of low-Cs cultivars, which can assure the stable production of agricultural products in the future.

References

Kobayashi NI, Nobori T (2014) New Food Ind 56:1–7 (in Japanese)

Kobayashi NI, Sugita R, Nobori T, Tanoi K, Nakanishi TM (2015, accepted) Tracer experiment using $^{42}K^{+}$ and $^{137}Cs^{+}$ revealed the different transport rates of potassium and caesium within rice roots. Funct Plant Biol.

Nobori T, Kobayashi NI, Tanoi K, Nakanishi TM (2014) Effects of potassium in reducing the radiocesium translocation to grain in rice. Soil Sci Plant Nutr 60:772–781

Qi Z, Hampton CR, Shin R, Barkla BJ, White PJ, Schachtman DP (2008) The high affinity K^{+} transporter AtHAK5 plays a physiological role *in planta* at very low K^{+} concentrations and provides a caesium uptake pathway in *Arabidopsis*. J Exp Bot 59:595–607

Robison WL, Brown PH, Stone EL, Hamilton TF, Conrado CL, Kehl S (2009) Distribution and ratios of ^{137}Cs and K in control and K-treated coconut trees at Bikini Island where nuclear test fallout occurred: effects and implications. J Environ Radioact 100:76–83

Rodriguez-Navarro A (2000) Potassium transport in fungi and plants. Biochim Biophys Acta 1469:1–30

Tanoi K, Kobayashi NI, Ono Y, Fujimura S, Nakanishi TM (2013) Radiocesium distribution in rice plants grown in the contaminated soil in Fukushima prefecture in 2011. Radioisotopes 62:25–29 (in Japanese)

Zhi Y-G, Shaw G, Nisbet AF, Wilkins BT (2002) Effect of external potassium supply and plant age on the uptake of radiocesium (^{137}Cs) by broad bean (*Vicia faba*): interpretation of results from a large-scale hydroponic study. Environ Exp Bot 47:173–187

Chapter 6
Consecutive Field Trials of Rice Cultivation in Partially Decontaminated Paddy Fields to Reduce Radiocesium Absorption in the Iitate Village in Fukushima Prefecture

Ichio Ii and Keitaro Tanoi

Abstract We performed consecutive field trials of rice cultivation to reduce radiocesium (^{134}Cs and ^{137}Cs) absorption by rice in a partially decontaminated paddy soil in the Iitate Village in Fukushima prefecture, Japan. People had evacuated this area because of the high levels of radioactive contamination caused by the nuclear disaster in 2011 at the Fukushima Daiichi Nuclear Power Station, Tokyo Electric Power. The radiocesium concentrations were measured for paddy soil and for lowland rice grown on variously decontaminated paddy soil in 2012 and 2013. The results show that the radiocesium concentration in the brown rice cultured in the fields of Sasu and Maeda with 2000–6000 Bq/kg dry weight (0–15 cm average soil depth) was below 40 Bq/kg, which is below the Japanese new standard for food (100 Bq/kg). In addition, the radiocesium concentration in the brown rice depended on the decontamination level of the paddy soil. In addition, the radiocesium concentration in the rice was reduced depending on the exchangeable K content of the soil, which plateaued around 20 mg K_2O per 100 g dry soil. However, in 2013, in a test field of Komiya where the radiocesium concentrations were higher than 8000 Bq/kg dry weight, brown rice with more than 100 Bq/kg was harvested, indicating the need for further decontamination. Overall, our results show that decontamination and additional K fertilization can reduce the radiocesium concentration in rice to less than the new standard, and that we could resume rice cultivation in the Iitate village by rather practical way.

Keywords Radioactive fallout • Soil • Rice • Tokyo electric power • Fukushima Daiichi nuclear accident • Cesium134 • Cesium137

I. Ii (✉)
NPO Resurrection of Fukushima, 1-3-6-2 F1, Asagayakita, Suginami-ku, Tokyo 166-0004, Japan
e-mail: Iii1234@chive.ocn.ne.jp

K. Tanoi
Graduate School of Agricultural and Life Sciences, The University of Tokyo, 1-1-1, Yayoi, Bunkyo-ku, Tokyo 113-8657, Japan

T.M. Nakanishi, K. Tanoi (eds.), *Agricultural Implications of the Fukushima Nuclear Accident*, DOI 10.1007/978-4-431-55828-6_6

6.1 Introduction

Rice is the most important crop in Japan. The Fukushima Daiichi nuclear accident in March 2011 caused radioactive material to spread out and down onto farm lands from the Fukushima Daiichi Nuclear Power Plant of Tokyo Electric Power, particularly in Fukushima prefecture. In the Iitate Village, Fukushima prefecture, people are still prohibited to live at home and to cultivate rice on their farms. It is an important issue for them to judge whether they can resume rice cultivation in the village within the next few years. Radiocesium (^{134}Cs and ^{137}Cs) is assumed to be the radioactive material of most concern at present, considering the quantity released and its half-life. In 1997, Tsukada et al. (2002, 2005) performed experimental cultivation of rice in the Rokkasyo Village in Aomori prefecture and measured ^{137}Cs in the soil and in rice parts. The ^{137}Cs level for the dry soil was 4.4 Bq/kg and the most was assumed to have derived from the many atmospheric atomic bomb experiments performed during the 1950s and 1960s. They reported that the ^{137}Cs radioactivity in dry white rice was 0.0048 Bq/kg. From these values, the transfer factor of ^{137}Cs for white rice from soil was calculated to be 0.0011. Uchida and Tagami (2007) reported the transfer factors of ^{137}Cs for brown rice cultivated in all of Japan to be 0.003–0.06 before the Fukushima disaster in 2011.

After the disaster, Ohmori et al. (2014) examined the effect of fertilization on the radiocesium concentration in rice grown in a paddy field in Kawamata-cho in Fukushima prefecture, and they revealed that excess N and K deficiency increased radiocesium accumulation in rice from the soil. Nobori et al. (2014) investigated the effect of K on the behavior of ^{137}Cs in hydroponically cultured rice plants and showed the importance of maintaining an appropriate K concentration before ear emergence to avoid ^{137}Cs contamination of the rice grains. In addition, in the confirmatory fieldwork, the Ministry of Agriculture, Forestry and Fisheries (MAFF 2012) reported that the radiocesium concentrations of brown rice cultivated in decontaminated paddy soils (Komiya and Kusanomukaioshi districts in the Iitate Village) were below 13 Bq/kg, but their relationship to the radiocesium concentrations in the soil was not clarified. Furthermore, the Fukushima prefecture agency and MAFF (in January 2013) and MAFF et al. (in March 2014) investigated the relationships between the radiocesium concentrations in brown rice and soil and the exchangeable K concentration in the soil and showed that the radiocesium concentrations in brown rice are negatively related to the concentrations of exchangeable K in soil. However, they also concluded that no clear relationship was observed between the radiocesium concentrations in brown rice and those of the soils where the rice was cultivated. Thus, we investigated the relationship between the radiocesium concentrations in the soil and in brown rice grown in variously decontaminated soils in test fields with and without K addition.

The approved specified NPO "Resurrection of Fukushima" (ROF; www.fukushima-saisei.jp/) is a volunteer organization that aims to rebuild lives and reconstruct agriculture-centered industries that have been affected by the nuclear power plant accident. The activities started by focusing on the Iitate Village in 2011

Fig. 6.1 The first visit to Iitate Village by volunteers in June 2011. A picture of Mr. Muneo Kanno with the visitors in front of his house in Sasu against the background of his bull barn and a stream. This visit provided the momentum to establish NPO "Resurrection of Fukushima" to work with villagers in the Iitate Village

in collaboration with the villagers (Fig. 6.1) and with assistance from the Graduate School of Agricultural and Life Sciences, University of Tokyo. The collaboration was facilitated by letter exchanges between the village head and the director of the faculty. One of the most important activities was that we annually cultivated rice in 2012, 2013, and 2014 in the Iitate Village, focusing on radiocesium in brown rice in relation to the level of decontamination of soil cultivated, with or without the addition of KCl fertilizer.

6.2 Experimental Rice Cultivation Procedures in 2012

Figure 6.2 shows diagrams and pictures of the test fields in the Iitate Village in 2012: (a) Sasu test fields (N37°44′, E140°43′) in Sasu district and (b) Maeda test fields (N37°43′, E140°40′) in Maeda district. The Sasu test fields were partially decontaminated in April 2012 by irrigating a paddy field with water to approximately 5-cm depth with rotary weeding tools, and the muddy water was then swept out (Mizoguchi 2013, Resurrection of Fukushima 2012).

The Sasu test fields were divided into fields A, B, C1, C2, and D, depending on the extent of decontamination. Each test field was divided into N (without additional KCl fertilizer) and K (with additional KCl fertilizer). The decontamination treatments were performed as follows:

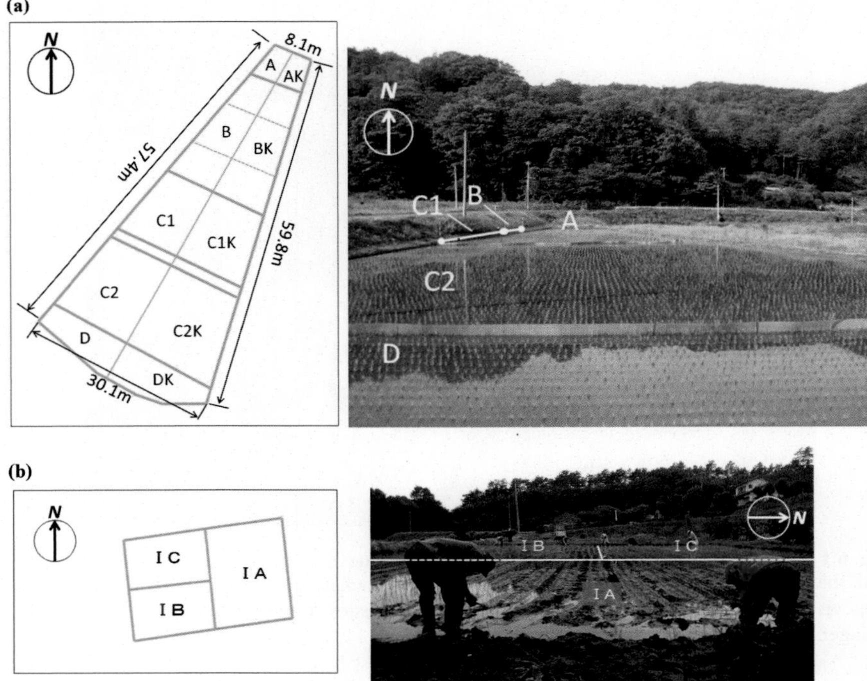

Fig. 6.2 The test fields of Sasu (**a**) and Maeda (**b**) in 2012 (Ii et al. 2015). (**a**) *Left*: diagram of test fields of Sasu. *Right*: picture of Sasu test fields just after the rice planting in June 2012. (**b**) *Left*: diagram of test fields of Maeda. *Right*: picture of Maeda test fields while rice planting in June 2012

A: Three cycles of shallow irrigation with rotary weeding tools and then drainage with a tennis court brush (0.5 acres)
B: One cycle of shallow irrigation with rotary weeding tools and then drainage with a tennis court brush (3.3 acres)
C1: Two cycles of shallow irrigation with a rotary weeding machine and then natural drainage (1.6 acres)
C2: One cycle of shallow irrigation with a rotary weeding machine and then natural drainage (5.2 acres)
D: No shallow irrigation (1.4 acres).

The Maeda test fields were divided into IA, IB, and IC with plowing and irrigating an approximately 15-cm depth soil.

Rice seedlings of Akitakomachi were planted in Sasu and Maeda fields as shown in Fig. 6.2. Before planting, basal fertilizer (12N:18P:16K:4Mg; weight % as N, P_2O_5, K_2O, and MgO) was mixed with the plowed soil at 40 kg per 10 acres in Sasu test fields. In Maeda test fields, basal fertilizer (10N:8P:8K:2Mg) was mixed with the plowed soil at 40 kg per 10 acres. The K fertilizer was added as KCl (20 kg per

10 acres) to K test fields with the basal fertilizer in Sasu. No KCl fertilizer was added to the Maeda test fields. No fertilizer was added after planting the rice.

Water was introduced to the test fields from a brook; the bottom water was blocked from entering. On rainy days, water was blocked from entering the Sasu test fields. For the Maeda test fields, bags of absorbent (Zeolite) were set at water entry points. The herbicide Sornet (Syngenta) was applied 1 week after rice planting. Electric nets were set to protect the fields from boar and monkey damage. Rice was harvested in mid-October in 2012. Figure 6.3 shows the test fields of Sasu (a) and Maeda(b) in October.

Fig. 6.3 (**a**) Sasu test fields with the sign in Japanese showing the rice cultivation trial on going by ROF. (**b**)Maeda test field at the mature stage in October 2012

Rice sampling was performed at the dough stage on September 15–16 and at the mature stage on October 6–7. Following a five-point sampling procedure, five sampling points were assigned to each test field, and 10–20 sheaves of rice plant were then cut and collected (Fig. 6.4a). The bundles of rice were dried indoors for more than 1 week, and then the rice bundles from each test point were collected and threshed with an old-fashioned thresher (Fig. 6.4b) to give one unhulled rice sample for each test field. The unhulled rice was sent to "Circle Madei"(Fig. 6.5), a volunteer employee and student group at Tokyo University that collaborates with

Fig. 6.4 (**a**) Sampling of sheaves of rice plants in test fields in Sasu in October 2012. (**b**) Threshing of rice bundles with an old-fashioned thresher to give unhulled rice in November 2012

(a)

福島県飯舘村の再生を応援します

サークル

「までい」(仮称)

福島県飯舘村は 2011 年 3 月の福島第1原子力発電所事故により、村民は全員避難生活を余儀なくされている。事故直後から多くのボランティアの人たちが、村の生活を再生させ村民の一日も早い帰村を目指して、毎週末に村民と一緒に緻密な放射能汚染の測定や農水省のマニュアルにない除染方法などを暗中模索しながら試験を行っている。しかしながら、試験結果の解釈には専門家のアドバイスを必要としている。

こうした状況の中、飯舘村で実施されているボランティアの試験結果の解釈に関して、9 月下旬に飯舘村村長から農学生命科学研究科長宛に研究協力要請の書簡が届き、研究科長もこの要請に受ける書簡を返した。これにより農学生命科学研究科の関係者が飯舘村で活動できるようになった。しかしながら、研究科内には何らかのボランティア活動に参加したいと思いながらも現地に足を運ぶことのできない教職員も多く存在する。

そこで、現地に行かなくても東京に居ながら飯舘村再生のボランティア活動に参加できる教職員サークルを立ち上げた。このサークルには農学生命科学研究科の教職員であれば誰でも参加できる。当面の活動は、飯舘村で除染した水田に試験栽培した東京地区のボランティアの方々と一緒に米の放射線量を測るための籾を玄米に仕上げる作業を行う予定である。

なお、サークル「までい（真手い）」とは、丁寧にという意味の飯舘村の方言からとった名称である。

(b)

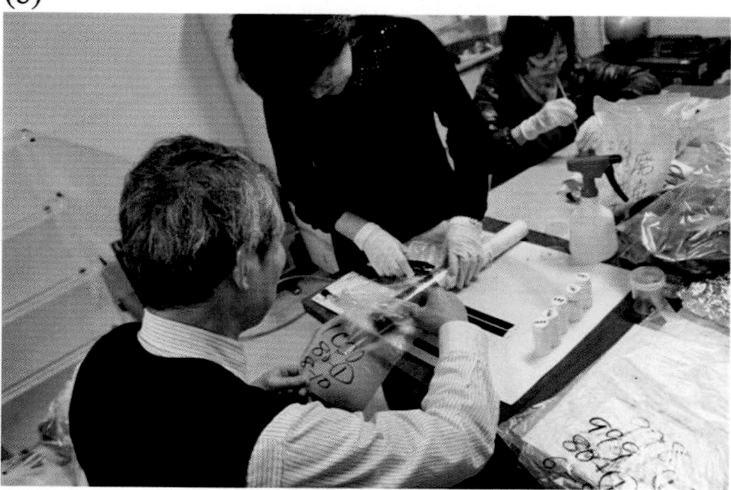

Fig. 6.5 (**a**) The poster of "Circle Madei" on the front of the circle room in the Graduate School of Agricultural and Life Sciences, University of Tokyo, showing its motivation in Japanese to support Iitate villagers to reconstruct their "Madei" life. "Madei" means polite, earnest, and steady, even if

(a)

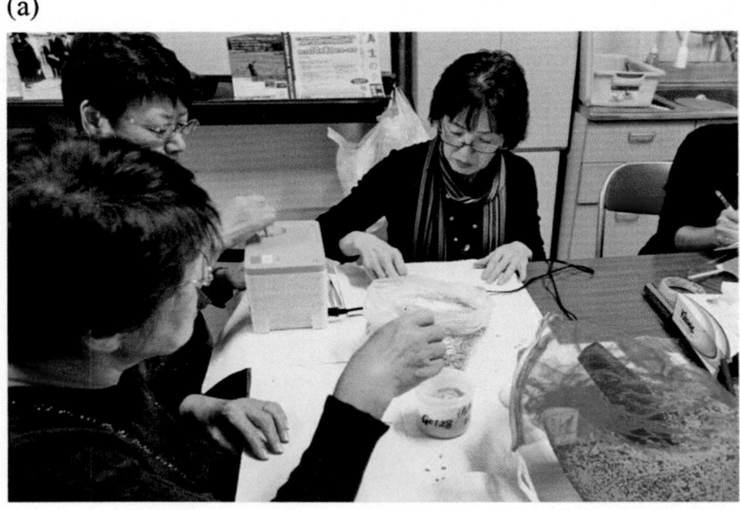

(b)

Fig. 6.6 (**a**) Hulling process of unhulled rice to give brown rice, being performed at "Circle Madei". (**b**) Polishing process of brown rice to give white rice and rice bran, being performed at "Circle Madei"

ROF to prepare samples for radioisotope measurement. The brown rice was prepared from the unhulled rice, using a hulling machine (Fig. 6.6a) and white rice and rice bran were prepared from the brown rice, using a rice polisher (Fig. 6.6b). The

---

Fig. 6.5 (continued) it may be slow (in dialect in Iitate Village, where they advocate "Madei" life). (**b**) Members of "Circle Madei", showing soil sample preparation No. 1000 in March 2013, since the start of November 2012. The number of samples reaching 10,000 in June 2015

radiocesium measurements were performed at the Radioisotope Center, using a Ge semiconductor detector (GEM and GMX type; Seiko EG&G) for rice samples in 250 ml containers, and a NaI (Tl) Scintillation counter (2480WIZARD2 Autoγcounter; Perkin Elmer) for soil samples in 20 ml vials (Nobori et al. 2013). The value of the soil was corrected per dry weight by measuring the soil weight after drying the soil at 60 °C for more than 6 days.

6.3 Results of Experimental Cultivation in 2012

6.3.1 Radiocesium Concentrations of Brown Rice and Soil

Figure 6.7 shows pictures of brown rice grains prepared at the dough stage and at the mature stage. Grains from the mature stage look browner than those from the dough stage. Figure 6.8 shows the radiocesium concentration of brown rice sampled from each test field at the dough stage and at the mature stage, respectively. Figure 6.9 shows the radiocesium concentration of soil (0–15 cm average depth) for each test field. The radiocesium transfer factor for each stage at each test field is shown in Fig. 6.10 and was calculated from the data presented in Figs. 6.8 and 6.9. The brown rice measurements show that the radiocesium concentration from either stage was less than 40 Bq/kg. In the KCl fertilized fields (AK, BK, C1K, C2K, and DK), the concentrations were below 25 Bq/kg. The transfer factors for brown rice were 0.002–0.008, whereas those in the KCl fertilized field were 0.002–0.003 at the dough stage and 0.003–0.004 at the mature stage.

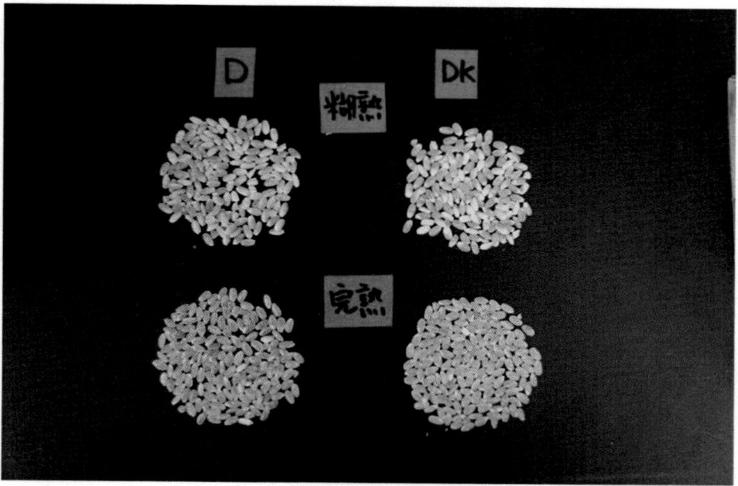

Fig. 6.7 Grains of brown rice prepared at the dough stage (*upper*) and at the mature stage (*below*) of test field D (*left*) and DK (*right*) in Fig. 6.2

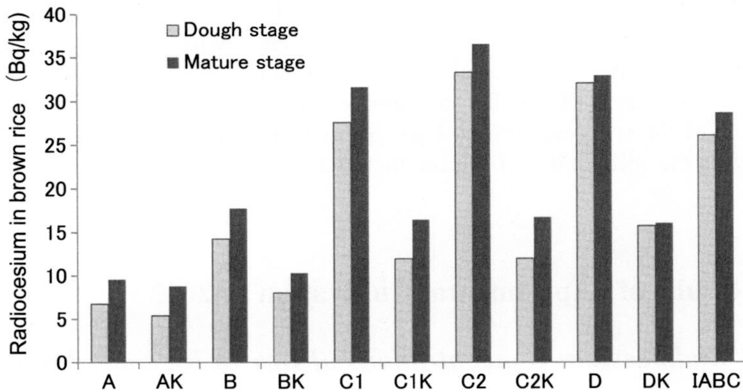

Fig. 6.8 The concentration of radiocesium in brown rice cultivated in Sasu (A–DK) and Maeda (IABC) in 2012 (Ii et al. 2015). Test fields are shown in Fig. 6.2

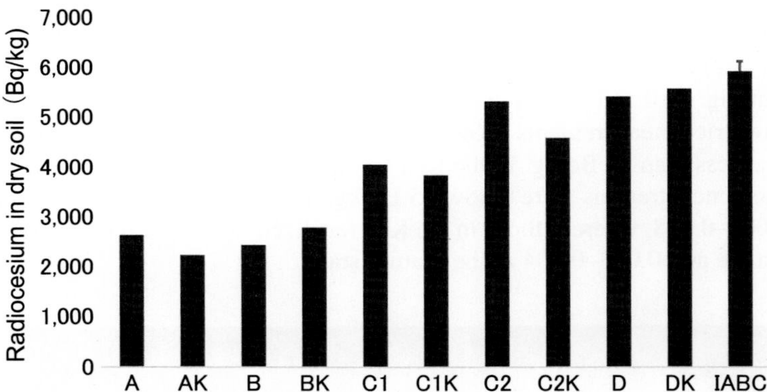

Fig. 6.9 The concentration of radiocesium in the soil (15 cm average depth) in 2012 (Ii et al. 2015). Test fields are shown in Fig. 6.2. The *bar* for IABC shows the SD of three soil samples (IA, IB, and IC)

Figure 6.11 shows the relationship between the radiocesium concentration in brown rice at the mature stage and that of the soil cultivated in the test field of Sasu for each group with KCl fertilizer (K) and with no KCl added (N). In both groups, the radiocesium concentration of brown rice tended to increase as the radiocesium concentration in the soil increased. A significant difference test of the correlation coefficient shows a significant p-value (0.032) for a group with only basal fertilizer added, but a non-significant p-value (0.057) for a group with KCl fertilizer added (at the significant test level of 0.05). The inhibitory effect of KCl fertilizer on radiocesium in brown rice was clear ($p = 0.026$; t-test, n = 5). The radiocesium concentration in brown rice at the mature stage was approximately 10 % higher than that from the dough stage (Fig. 6.8). The reason for this is unknown.

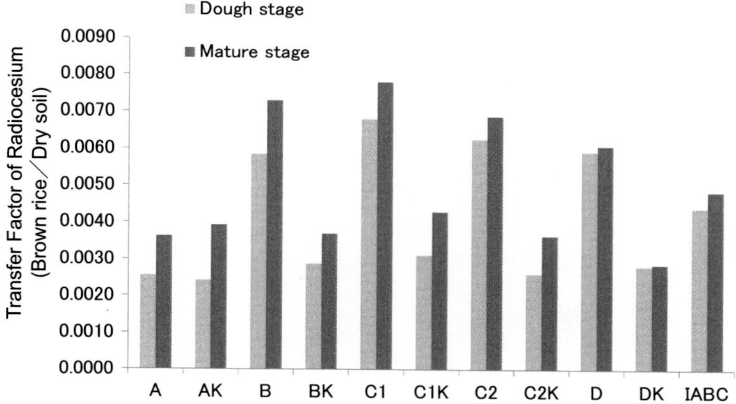

Fig. 6.10 The transfer factor of brown rice in 2012 calculated from the data shown in Figs. 6.8 and 6.9 (Ii et al. 2015). The test fields are shown in Fig. 6.2

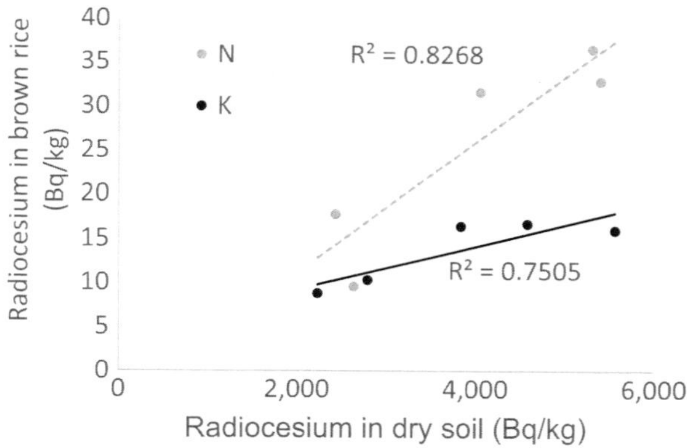

Fig. 6.11 The relationship of the radiocesium concentration between soil and brown rice cultivated in the test fields of Sasu in 2012 (Ii et al. 2015). The *p*-value between the two groups (K: with KCl fertilizer; N: no KCl fertilizer) was 0.026 (*t*-test, n = 5)

6.3.2 Analyses of Exchangeable Cations in Soil from the Test Fields

The soils measured for radiocesium were also analyzed for exchangeable cations using ICP-OES (Optima 7300DV) after extracting the dry soil with 1 M ammonium acetate solution at room temperature for 24 h. The results are shown in Fig. 6.12. The fields with added KCl (AK, BK, C1K, C2K, and DK) had higher exchangeable K than the corresponding fields without KCl addition (A, B, C1, C2, and D), respectively. However, there were significant differences between the test fields; for example, the exchangeable K content of A was much higher than for B, C1,

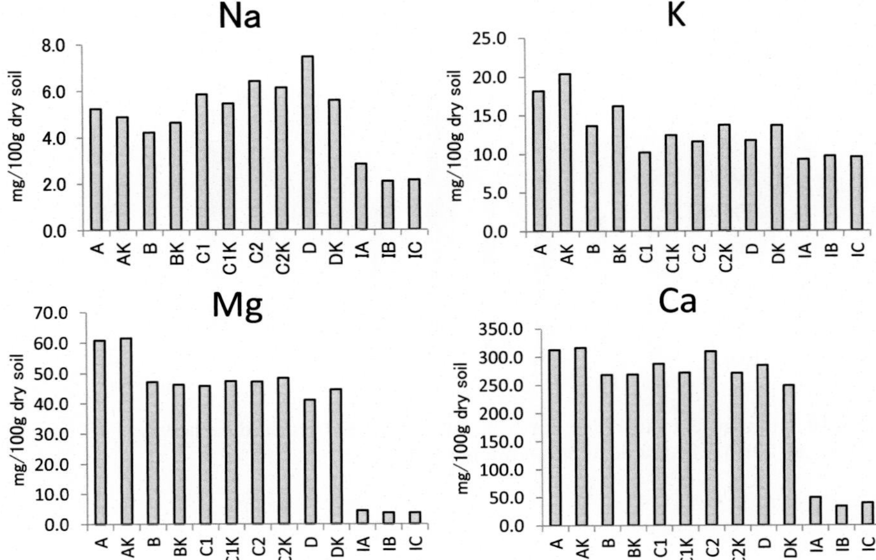

Fig. 6.12 Exchangeable cations of soil in Sasu and Maeda test fields in 2012 (Ii et al. 2015). Exchangeable Na, K, Mg, and Ca in dry soil are expressed as mg per 100 g of dry soil (*vertical axis*)

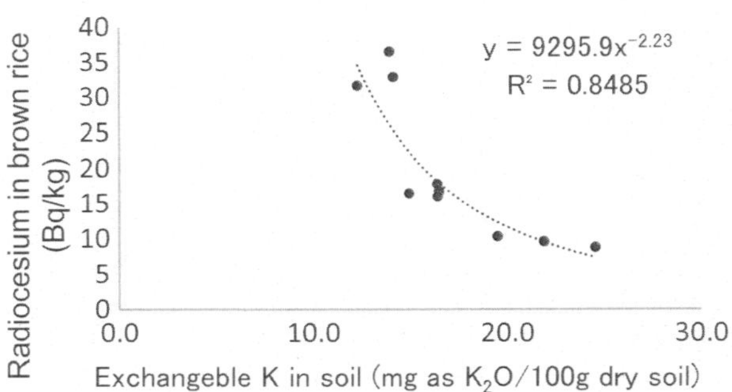

Fig. 6.13 Relationship of the radiocesium concentration in brown rice to exchangeable K in soil in Sasu test fields in 2012 (Ii et al. 2015). Exchangeable K is expressed in mg of K_2O per 100 g of dry soil

C2, and D. All measured cation contents in the test fields of Maeda were much lower than those of Sasu. This is assumed to be because of the much higher sand content in the Madea field soil (IA, IB, and IC). This is also considered to be one of the causes of low rice yields in the Maeda test fields. Figure 6.13 shows the relationship between the radiocesium concentration in brown rice at the mature stage and exchangeable K in the soil in the Sasu test fields. A higher exchangeable K resulted in a significantly lower radiocesium concentration in cultivated brown rice. The radiocesium concentration in brown rice was below 10 Bq/kg when the exchangeable K content was higher than 20 mg/100 g dry soil (mg as K_2O). This indicates that maintaining exchangeable K content in soil higher than 20 mg/100 g is extremely important for reducing the radiocesium concentration of cultivated brown rice. This is consistent with the directive of National Agriculture and Food Organization (NARO 2012). The transfer factor of radiocesium in brown rice to that in soil was 0.003–0.004 when the exchangeable K content in soil was higher than 20 mg/100 g. This is consistent with the lower range of reported values by Uchida and Tagami (2007).

6.3.3 Radiocesium Concentration in White Rice and in Rice Bran

Figure 6.14 shows the radiocesium concentrations in white rice and in rice bran prepared from the brown rice from each test field. The concentration in white rice was below half the value of brown rice and all were lower than 10 Bq/kg. The radiocesium concentrations in rice bran were more than double than that in brown rice. The values in some fields (C2, D, and IABC) were higher than 100 Bq/kg, although others harvested from partially decontaminated fields (A, B, and C1) and from KCl added fields were below 100 Bq/kg (the new standard for food in April 2012).

Thus, the 2012 field trials suggest that rice cultivation in the Iitate Village is feasible by reducing soil radiocesium by decontamination using shallow soil mixing and drainage, and by addition of KCl together with basal fertilizer.

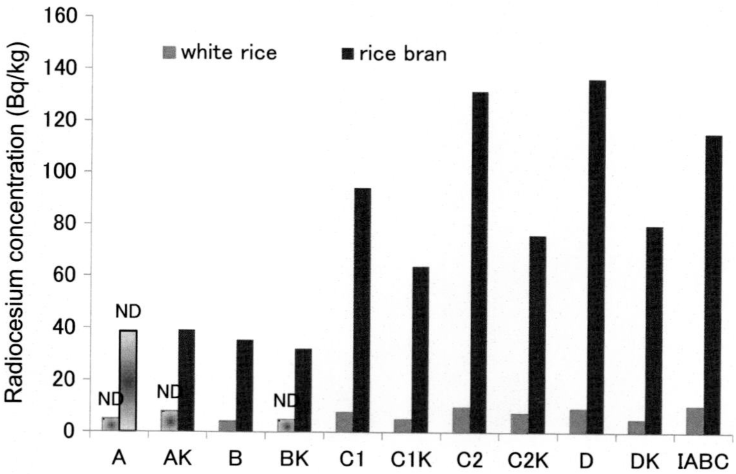

Fig. 6.14 Radiocesium concentration in white rice (*left column*) and rice bran (*right column*) harvested in each test field in 2012 (Ii et al. 2015). ND shows that both ^{134}Cs and ^{137}Cs are below the detection limit. The height of each ND column shows the sum of the detection limits of ^{134}Cs and ^{137}Cs

6.4 Experimental Cultivation of Rice in 2013

In addition to Sasu and Maeda test fields in 2012, an experimental cultivation was performed in Komiya test fields (N37°37′, E140°46′) in Komiya district (Fig. 6.15a, b) and in "Sasu Madei" test fields located to the eastern side of the

Fig. 6.15 (**a**) Komiya test fields in 2013, with the sign in Japanese showing that test cultivation was in progress. (**b**) Komiya test fields just after rice planting in August 2013. (**c**) "Sasu Madei" test fields just after rice planting in 2013

Table 6.1 Description of test paddy fields in 2013. N: K not added, K: KCl added (Ii et al. 2015)

Farm	Decontamination method (year)	Test field name	KCl added or not
Komiya	**Deep irrigation (ca.15 cm depth, 2013)**	**S3 3** cycles	**N, K**
		S1 1 cycle	**N, K**
		S0 0 cycle	**N, K**
Maeda	**Deep irrigation (2012)**		**N, K**
Sasu	**Shallow irrigation (ca.5 cm depth, 2012)**	**AB**: A and a part of B in 2012	**N, K**
		WC: a part of B and C1 in 2012 with straw plowed in the soil	**N. K**
		CD: C2 and D in 2012	**N, K**
Sasu Madei	**Madei method (ca.5 cm depth of surface soil removed, 2013)**	**W**: Straw plowed in	**N, K**
		N: No straw plowed in	**N, K**
		T: Compost mixed in	**N, K**
		MK: KCl added with basal fertilizer (8 in June)	**K**
		YK: KCl added during the panicle formation stage (28 in July)	**K**
		HK: KCl added during the heading stage (16 in August)	**K**

Sasu test fields in 2012 (Fig. 6.15c). Table 6.1 shows a list of the test fields, treatment methods, test field names, and whether KCl was added or not. The straw harvested in 2012 (approximately 80 kg/acre) was plowed into some of the test fields (WC and W) because straw contains high K and other nutrients and is generally recycled into the soil after rice harvest to keep the soil fertilized.

Rice cultivation at Sasu and Maeda was performed in a similar way to that in 2012, using Akitakomachi rice seedlings with some modifications. Briefly, basal fertilizer 12N:18P:16K:4Mg (40 kg per 10 acres) was applied to all the test fields and KCl (20 kg per 10 acres) was applied to the test fields affixed with K in early June. Rice was planted in mid-June (Fig. 6.16) and then harvested in mid-October (Fig. 6.17). In the Komiya test fields, basal fertilizer 12N:18P:16K:4Mg (40 kg per 10 acres) was added in mid-May. Rice planting and KCl (20 kg per 10 acres) addition to K affixed test fields was performed in late May. The rice was harvested in early October. Rice sampling at the mature stage was performed at Sasu and Maeda in mid-October and at Komiya in early October. Measurements of radiocesium were performed the same way as those in 2012, except that rice bran was measured using an NaI(Tl) scintillation counter with 20 ml vials.

Fig. 6.16 (**a**) Maeda test field showing rice planting by members of ROF in June 2013. (**b**) Lunch to celebrate rice planting at the field close to Sasu test field in June 2012. After rice planting, we usually have dinner with sake (rice wine) to celebrate rice planting and pray for safe growth of rice with good weather

(a)

(b)

Fig. 6.17 (**a**) Havested rice hung for natural drying at Sasu in October 2013. (**b**) Lunch to celebrate rice harvest at Sasu in October 2012. Rice harvest is a most important event in farmer life in Japan

6.5 Results of Experimental Cultivation in 2013

6.5.1 Radiocesium Concentrations of Brown Rice and Soil

Figure 6.18 shows the radiocesium concentrations of brown rice sampled at the mature stage and Fig. 6.19 shows the radiocesium concentrations of soil (0–15 cm average depth) for each test field. The radiocesium transfer factor for each test field is shown in Fig. 6.20 and was calculated from the data in Figs. 6.18 and 6.19. Soil sampling at Komiya was performed from N sections and K sections together. In the test field of Maeda, soil sampling was not performed. In Komiya test fields, the radiocesium concentrations in the soil were high. This is due to high radiocesium concentrations before decontamination (around 14,000 Bq/kg dry soil measured in May 2013) and because the decontamination method of deep irrigation was not as effective as the shallow irrigation method performed at Sasu test fields in 2012. Furthermore, the differences in the radiocesium concentrations between S0, S1, and

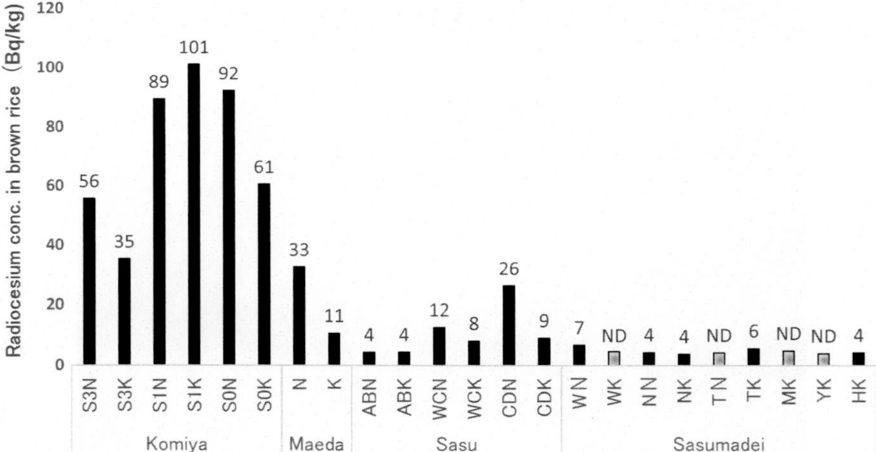

Fig. 6.18 The concentration of radiocesium in brown rice in 2013 (Ii et al 2015). The last letter "N" indicates the field "not KCl added" and the last letter "K" indicates the field "KCl added". Komiya: test fields of Komiya district (S3: decontaminated in 2013 by three cycles of mixing surface soil of approximately 15-cm depth and drainage; S1: the same as S3 except by one cycle of mixing; S0: the same as S3 except that there was no cycle of mixing). Maeda: test fields of Maeda district. Sasu: Sasu test fields of Sasu district decontaminated by rotary weeding method of mixing the surface soil to about 5-cm depth and drainage in 2012 (AB: A plus a part of B test field in 2012; WC: A part of B plus C1 test field in 2012 and straw harvested in 2012 was mixed in the soil; CD: C2 plus D in 2012). Sasumadei: Sasu test fields decontaminated by stripping approximately 5 cm of surface soil in 2013 (WN: straw was mixed in the soil, no KCl added; WK: straw was mixed in and KCl added; NN: no straw mixed in, no KCl added; NK: no straw mixed in, KCl added; TN: compost was mixed in, no KCl added; TK: compost was mixed in, KCl added; MK: KCl added with base fertilizer; YK: KCl added during the panicle formation stage; HK: KCl added during the heading stage). ND shows that both ^{134}Cs and ^{137}Cs are below the detection limit. The height of the ND columns show the sum of the detection limits of ^{134}Cs and ^{137}Cs

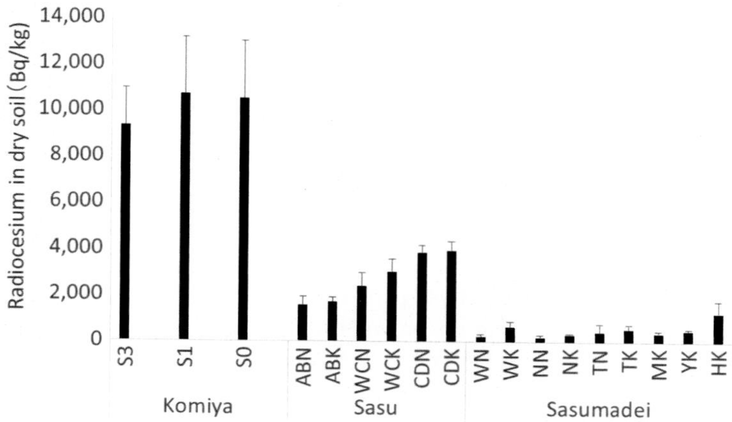

Fig. 6.19 The concentration of radiocesium in dry soil (15-cm average depth) in 2013 (Ii et al. 2015). The *columns* and *bars* show average and SD, respectively, of values of five soil points in each test field. Refer to Fig. 6.18 legend for details of the test fields

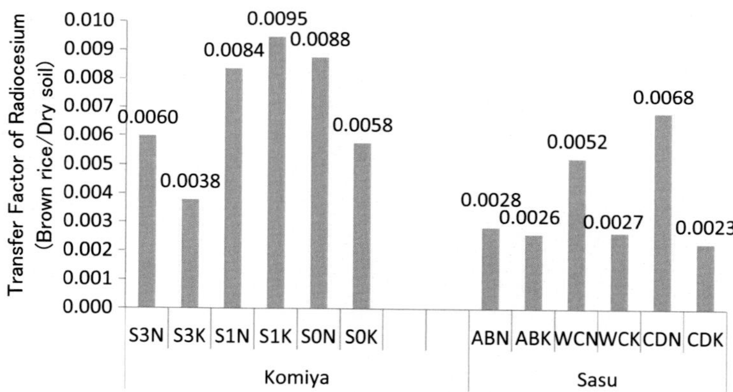

Fig. 6.20 Transfer factor of brown rice in 2013 calculated using the data shown in Figs. 6.18 and 6.19 (Ii et al. 2015). Refer to the legend of Fig. 6.18 for details of the test fields

S3 (Fig. 6.19) were less than the differences before planting the rice, which was probably due to soil transfer between the test fields and influx of radiocesium from the brook water (Nemoto and Abe 2013).

In the Sasu test fields, the radiocesium concentrations were 1500–4000 Bq/kg and in a similar order to those in 2012 (Fig. 6.9; ABN, ABK < WBN, WBK < CDN, CDK), although they were generally approximately 20 % lower than in 2012. No significant increase in the radiocesium concentrations occurred for the soil in the WBN and WBK treatments that contained the plowed straw harvested in 2012.

In "Sasu Madei" test fields, the radiocesium concentrations of the soils were 200–600 Bq/kg, except for a test field of HK (southern corner field), showing that the "Madei Method" of stripping co. 5 cm surface soil was effective at reducing the radiocesium concentration.

The radiocesium concentration of brown rice from the test fields in Komiya were 35–101 Bq/kg. The S3N and S3K fields showed lower values than the corresponding fields S0N, S0K and S1N, S1K. Addition of KCl reduced radiocesium accumulation in brown rice in S3 and S0, but not in S1. The reason for high radiocesium concentrations in S1K is unclear. In the Maeda test fields, there was a clear effect of added KCl (33–11 Bq/kg). In the fields of Sasu, ABN and ABK had 4 Bq/kg and WCB had 12 Bq/kg, whereas WCK had 8 Bq/kg. The CDN had 26 Bq/kg, whereas CDK had 9 Bq/kg. The reducing effect of KCl addition on the radiocesium concentrations in brown rice was observed in five test fields, excluding S1 in Komiya and AB. In "Sasu Madei", all were below 7 Bq/kg.

The transfer factor for brown rice from Komiya and Sasu was 0.002–0.01 (Fig. 6.8). The reducing effect of KCl addition was clear, except for Komiya S1 field.

6.5.2 Radiocesium Concentrations of White Rice and Rice Bran in 2013

Figure 6.21 shows the radiocesium concentrations in white rice and in rice bran prepared from the brown rice from Komiya and Maeda. The radiocesium concentration in white rice were approximately half of that of brown rice. White rice from

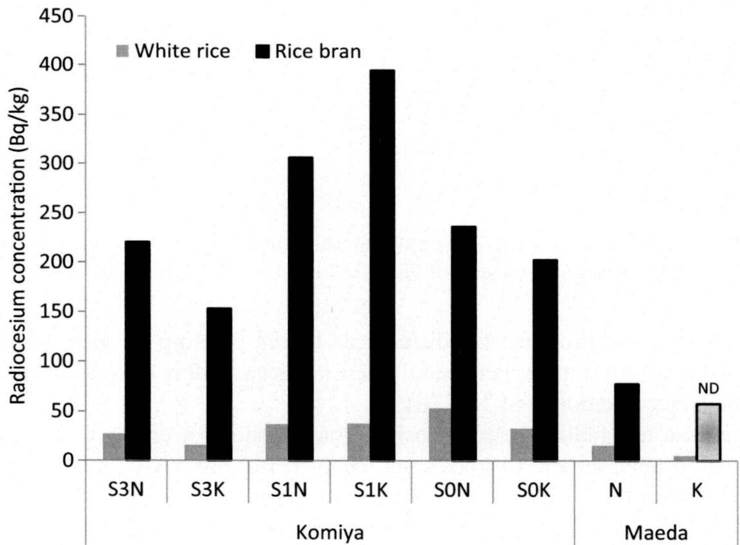

Fig. 6.21 Radiocesium concentration in white rice (*left column*) and rice bran (*right column*) harvested in each test field of Komiya and Maeda in 2013 (Ii et al. 2015). Refer to Fig. 6.18 legend for details of the test fields. ND shows that both ^{134}Cs and ^{137}Cs are below the detection limit. The height of the ND column shows the sum of the detection limits of ^{134}Cs and ^{137}Cs

Komiya had 15–53 Bq/kg, whereas white rice from Maeda had 15 Bq/kg in the field without KCl addition (N) and 5 Bq/kg in the field with KCl addition (K). White rice from CDN in Sasu was 10 Bq/kg, and white rice from the other test fields were below the detection limit. Rice bran from Maeda and Sasu both had a maximum of 77 Bq/kg, which is below the new standard for food; however, rice bran from Komiya exceeded this standard.

6.6 Conclusive Remarks

We performed field trials of rice cultivation in partially decontaminated paddy fields in the Iitate Village in Fukushima prefecture in 2012, 2013, and 2014. The results of 2012 and 2013 show that suitable decontamination and addition of KCl can reduce the radiocesium concentrations in brown rice to values much lower than the safety standard for food in Japan (100 Bq/kg). This trend is consistent with results in 2014, which are not detailed here. The rice harvested in 2012 and 2013 was not for consumption because it was a trial, even though it satisfied the standard levels. However, the members of ROF enjoyed eating the rice harvested at Sasu in 2014, after confirmation that the radiocesium concentration was below 25 Bq/kg based on both our measurement of rice samples and the results from the Fukushima prefecture screening test of all bags of brown rice harvested in 2014. The members of ROF will continue to cultivate rice in the Iitate Village with the villagers to rebuild lives and reconstruct agriculture-centered industries.

Acknowledgments "All rice fieldworks were performed with the cooperation of Mr. Muneo Kanno, Agricultural Committee in the Iitate Village, Mr. Ryuzo Ito and Mr. Kin-ichi Ookubo in the Iitate Village, Dr. Masaru Mizoguchi at the University of Tokyo, Dr. Junko Nishiwaki at the Ibaraki University, and members of the Resurrection of Fukushima" (representative: Mr. Yoichi Tao). Sincere appreciation to the members of ROF and " Circle Madei" at Tokyo University particularly to Ms. Tomiko Saito and Mr. Yoshio Uno, for sample preparation for radioisotope measurements, and to Dr. Natsuko I. Kobayashi, Dr. Atsushi Hirose, Dr. Naoto Nihei, and Mr. Tatsuya Nobori at the Graduate School of Agricultural and Life Sciences, The University of Tokyo, for radioisotope measurement and valuable comments. We sincerely thank Mr. Ryuzo Ito who cooperated with us for the use of his field for experimental cultivation and for the fieldwork, and who passed away on September 23, 2013. We sincerely acknowledge his kind cooperation during his life and pray for his repose.

Field trials of rice cultivation in 2012 were performed under co-research agreement with NARO. The research was entitled "Rice cultivation in test fields decontaminated by rotary weeding tools and confirmation of volume reduction and stabilization technology of the harvest."

References

Ii I, Tanoi KT, Uno Y, Nobori T, Hirose A, Kobayashi N, Nihei N, Ogawa T, Tao Y, Kanno M, Nishiwaki J, Mizoguchi M (2015) Radioactive cesium concentration of lowland rice grown in the decontaminated fields in Iitate-Village in Fukushima. Radioisotopes 64:299–310 (in Japanese with English summary)

Ministry of Agriculture, Forestry and Fisheries (2012) Results of test cultivation of agricultural products in the farm decontaminated (in Japanese) http://www.maff.go.jp/j/nousin/seko/josen/pdf/sakutuke.pdf. Accessed Nov 2012

Ministry of Agriculture, Forestry and Fisheries (MAFF), Fukushima Prefecture (2013) Analysis of the factors to cause rice of high radiocesium and counteractions (in Japanese) http://www.pref.fukushima.lg.jp/download/1/youinkaiseki-kome130124.pdf. Accessed Jan 2013

Ministry of Agriculture, Forestry and Fisheries (MAFF), Fukushima Prefecture, NARO, National Institute for Agro-Environmental Sciences (NIAES) (2014) Analysis of the factors to cause rice of high radiocesium and counteractions. Rev.2 (in Japanese) http://www.maff.go.jp/j/kanbo/joho/saigai/pdf/youin_kome2.pdf. Accessed Mar 2014

Mizoguchi M (2013) Remediation of paddy soil contaminated by radiocesium in Iitate Village in Fukushima prefecture. In: Nakanishi TM, Tanoi K (eds) Agricultural implications of the Fukushima nuclear accident. Springer, Japan, pp 131–142

National Agriculture and Food Research Organization (NARO) (2012) K fertilization to reduce radiocesium concentration (in Japanese) http://www.naro.affrc.go.jp/publicity_report/press/laboratory/narc/027913.html Accessed Feb 2012

Nemoto K, Abe J (2013) Radiocesium adsorption by rice in paddy field ecosystems. In: Nakanishi TM, Tanoi K (eds) Agricultural implications of the Fukushima nuclear accident. Springer, Tokyo, pp 19–27

Nobori T, Tanoi K, Nakanishi T (2013) Method of radiocesium determination in soil and crops using a NaI(Tl) scintillation counter attached with a autosampler. Jpn J Soil Sci Plant Nutr (in Japanese) 84:182–186

Nobori T, Kobayashi NI, Tanoi K, Nakanishi TM (2014) Effects of potassium in reducing the radiocesium translocation to grain in rice. Soil Sci Plant Nutr 60:772–781

Ohmori Y, Kajikawa M, Nishida S, Tanaka N, Kobayashi NI, Tanoi K, Furukawa J, Fujiwara T (2014) The effect of fertilization on cesium concentration of rice grown in a paddy field in Fukushima prefecture in 2011 and 2012. J Plant Res 127:67–71

Resurrection of Fukushima (2012) Activity record in April, 2012. http://www.fukushima-saisei.jp/archives/report201204.html. Accessed Apr 2012

Tsukada H, Hasegawa H, Hisamatsu S, Yamasaki S (2002) Rice uptake and distributions of radioactive ^{137}Cs, stable ^{133}Cs and K from soil. Environ Pollut 117:403–409

Tsukada H, Takeda H, Takahashi T, Hasegawa H, Hisamatsu S, Inaba J (2005) Uptake and distribution of 90Sr and stable Sr in rice plants. J Environ Radioact 81:221–231

Uchida S, Tagami K (2007) Soil-to-plant transfer factors of fallout ^{137}Cs and native ^{133}Cs in various crops collected in Japan. J Radioanal Nucl Chem 273:205–210

Chapter 7
Effects of "Clean Feeding" Management on Livestock Products Contaminated with Radioactive Cesium Due to the Fukushima Daiichi Nuclear Power Plant Accident

Noboru Manabe, Tomotsugu Takahashi, Maiko Endo, Chunxiang Piao, Junyou Li, Hiroshi Kokado, Minoru Ohta, Keitaro Tanoi, and Tomoko M. Nakanishi

Abstract Contamination of food and animal products by radioactive cesium represents an important potential route of exposure in the human food chain. Therefore, following the Fukushima Daiichi nuclear power plant accident, the development of solutions for radiocesium contamination is a serious social issue in Japan. Most farm animals are kept in closed barns in Japan; this reduced the initial contamination of animal products by radioactive nuclides in the early phase of the accident. Furthermore, pigs and chickens were given imported feed that was not contaminated. However, more than 10 million tons of grass feed were needed for herbivorous livestock in Japan. We report the effects of "clean feeding" management on livestock products contaminated with radioactive cesium due to the nuclear accident. The present results concerning products from herbivores (horses, sheep, and cattle) revealed that radiocesium levels were undetectable in the meat or milk of herbivores that fed on non-contaminated feed after an initial period of consuming radiocesium-contaminated feed. Thus, we conclude that

N. Manabe (✉)
Animal Resource Science Center, Graduate School of Agricultural and Life Sciences, The University of Tokyo, Kasama 319-0206, Japan

Osaka International University, Osaka 570-8555, Japan
e-mail: n-manabe@oiu.jp

T. Takahashi • M. Endo • C. Piao • J. Li
Animal Resource Science Center, Graduate School of Agricultural and Life Sciences, The University of Tokyo, Kasama 319-0206, Japan

H. Kokado • M. Ohta
Miho Training Center, Japan Racing Association, Ibaraki 300-0415, Japan

K. Tanoi • T.M. Nakanishi
Radioisotope Center, Graduate School of Agricultural and Life Sciences, The University of Tokyo, Tokyo 113-8657, Japan

T.M. Nakanishi, K. Tanoi (eds.), *Agricultural Implications of the Fukushima Nuclear Accident*, DOI 10.1007/978-4-431-55828-6_7

"clean feeding" management can play a crucial role in solving the social problem of food contamination.

Keywords Cattle • Clean feeding • Feeding management • Herbivorous livestock • Horses • Sheep • Radioactive cesium • The Fukushima Daiichi nuclear power plant accident

7.1 Introduction

On March 11th, 2011, the Japanese Prime Minister declared that a nuclear disaster occurred at Fukushima Daiichi nuclear power. This disaster resulted from an earthquake that occurred in Eastern Japan. The Japanese government subsequently issued various correspondences, and on March 17, 2011, the Ministry of Health, Labor and Welfare, Japan (MHLW) was notified of the handling of foods that were contaminated with radionuclides (hereafter referred to as radioactive cesium). They established provisional radionuclides reference/regulation levels for food and drinking water, under the guide of food sanitation law. Radioactive cesium levels (cesium-137 and cesium-134) in drinking water, milk, and dairy products had to be less than 200 Bq/kg, while those in other foods, including cereals, vegetables, fruits, meat, eggs, and fish had to be less than 500 Bq/kg. In response to the provisional reference/regulation values, the Ministry of Agriculture, Forestry and Fisheries, Japan (MAFF) negotiated a rule that prevented food contamination by radionuclides. On April 14, 2011, the MAFF established regulations to reduce radioactive cesium contamination in livestock and livestock products such as milk and meat. Radioactive cesium levels in feed for dairy cattle, beef cattle, horses, pigs and chickens had to be less than 300 and feed for farmed fish had to be less than 100 Bq/kg. The provisional reference/regulation values were subsequently evaluated from various angles, and radioactive cesium levels in human food and drinking water were significantly revised. New reference/regulation values were applied on April 1, 2012. Therefore, radioactive cesium levels had to be maintained at less than 10 Bq/kg in drinking water, less than 50 Bq/kg in milk and infant foods, and less than 100 Bq/kg in other foods (such as cereals, vegetables, fruits, meat, eggs, and fish). These reference/regulation values are still applicable today. In response to such revisions, on February 3, 2012 the MAFF also revised radioactive cesium levels in livestock feed and bedding, and fertilizers such as compost used in crop production. However, these revised values were considered insufficient to meet the new reference values for human foods. Therefore, the MAFF established further revisions on March 23, 2012, which were referred to as new tolerance values. With these new tolerance levels, radioactive cesium levels had to be less than 100 Bq/kg in feeds for cattle and horse, 80 Bq/kg in pig feed, 160 Bq/kg in chicken feed, and 40 Bq/kg feed for farmed fish. Moreover, radioactive cesium levels in the litters of livestock and fertilizer containing compost had to be less than 400 Bq/kg.

The reference/regulation values for human foods and tolerance values for animal feeds were established immediately after the Fukushima Daiichi nuclear power plant accident, and were based on overseas findings, not those in Japan. Farming systems in Japan differ from those in overseas countries. Japanese people have characteristic eating habit and farming systems; therefore, fundamental research that reflects the reality of these Japanese styles needs to be considered when these reference/regulation and tolerance values are revised.

After the nuclear power plant accident, researchers in Japan performed empirical research using livestock under conditions that reflected feed management in Japan in order to determine and subsequently alleviate radioactive cesium contamination levels in agricultural areas. We investigated how the levels of radioactive cesium migrated from feed to livestock and livestock products using farm animals under conditions that reflected the realities of feeding management in Japan in an attempt to maintain the health of the Japanese population. It is approximately 4 years since the nuclear power plant accident, and many of our research activities are ongoing. Although most of our findings are incomplete or are only provided as interim reports, we hope that the results described herein will contribute to the reconstruction of the livestock industry after the Fukushima Daiichi nuclear power plant accident.

7.2 Absorption and Accumulation of Radioactive Cesium from Feed to Horse Meat and the Effect of "Clean Feeding" Management

In mid-March, 2011, livestock feeds (such as grass and rice straw) were contaminated with radionuclides including radioactive cesium. Contamination occurred over wide areas of Tohoku and Kanto, with a radius of approximately 200 km from the Fukushima Daiichi nuclear power plant. However, the dynamics of radioactive cesium from contaminated feeds to farm animals have not yet been elucidated in detail. Radioactive cesium-contaminated horse meat was found in November, 2012. Radioactive cesium (115.6 Bq/kg) was detected in horse meat processed at a local slaughterhouse (Koriyama-city, Fukushima prefecture, Japan) approximately 60 km from the nuclear power plant at a level that slightly exceeded the new reference/regulation value (100 Bq/kg). Following this incident, we investigated the transition of radioactive cesium from feed to horse meat using contaminated grass haylage produced at an experimental ranch of the Animal Resource Science Center, the University of Tokyo (Kasama, Japan) in the spring of 2011. The fate of radioactive cesium in horse meat after initiating a diet of non-contaminated grass haylage was assessed.

7.2.1 Experimental Procedure

(A) Preparation of horse feed containing radioactive nuclides caused by the accident at the Fukushima Daiichi nuclear power plant: Grass (Italian ryegrass, *Lolium multiflorum* Lam.) was seeded in October 2010 and was cultivated in a grass field at the experimental ranch of the Animal Resource Science Center. After mowing between the 10th and 15th of May, 2011 (2 months after the accident), raw grass (radioactive iodine: undetectable levels, and radioactive cesium: 113 Bq/kg) was dried for several days and then packed into plastic film to prepare anaerobically fermented grass forage (haylage). The haylage contained radioactive nuclides (radioactive iodine: undetectable level and radioactive cesium: 480 Bq/kg).

(B) Horse feed without radioactive nuclides: we used haylage that was produced in May 2013 at the Animal Resource Science Center. This haylage contained undetectable levels of iodine or cesium. Horses were fed the non-contaminated haylage and mixture feed (0.5 kg/400 kg of body weight/day) purchased from JRA Facilities Co., Ltd. (Tokyo, Japan) depending on the physical condition of the horses.

(C) A study to assess the effects of "clean feeding" management: Six horses (male and female, 3–4 years old) that were kept in the stables of the Animal Resource Science Center were used. They were fed only non-contaminated haylage (10 kg/400 kg of body weight/day) for 4 weeks before being examined in June 2013 (380 ± 28 kg body weight at the end of pre-feeding, $n = 6$). These horses were then given contaminated haylage (10 kg/400 kg of body weight/day) for 8 weeks (404 ± 39 kg body weight at the end of feeding, $n = 6$) followed by "clean feeding" with non-contaminated haylage and mixture feed for 4 weeks (391 ± 38 kg body weight at the end of feeding, $n = 6$), 8 weeks (388 ± 13 kg body weight at the end of feeding, $n = 4$), and 16 weeks (371 ± 15 kg body weight at the end of feeding, $n = 2$). During the study period, each horse's feed intake was measured daily. The horses were given free access to water, which contained undetectable levels of iodine or cesium, through automated system. Following the administration of radioactive cesium-contaminated haylage, and 4 and 8 weeks after the initial administration of non-contaminated haylage, glutei skeletal muscle samples were collected by biopsy under sedation and local anesthesia. We sacrificed two horses each at 4, 8, and 16 weeks after the initial administration of non-contaminated haylage, under general deep anesthesia. We collected samples from the heart, spleen, liver, kidney, and psoas major muscle to estimate radioactive cesium contamination and conventional histopathology. Blood and feces were also collected. The radioactivity of iodine-131, cesium-134, and cesium-137 in each horse was measured biweekly. The body weight of each horse was measured every 4 weeks. During the experiment, a veterinarian diagnosed the health status of the horses, and conducted blood biochemical tests and hematological analyses using automatic analyzers every 4 weeks.

(D) Measurement of radioactive elements: The concentrations of radioactive elements in each sample were measured using a germanium semiconductor detector, and each nuclide was identified by gamma-ray spectrometry. Cesium-134 and cesium-137 were quantified at 661.6 and 604.7 keV, respectively, and each Bq value was calculated by calibrating the count value. Each radionuclide concentration was calculated based on the weight of each sample. The detection limit was set to three times the standard deviation of the background radiation level(s).

7.2.2 *Results*

(A) Body weight and feed intake: No significant differences were observed in body weight or feed intake of animals between the non-contaminated haylage feeding period and the contaminated haylage feeding period (data not shown).
(B) Health status: No serious symptoms were noted in any horse during the non-contaminated haylage feeding period or contaminated haylage feeding period. Furthermore, no significant differences were observed in blood biochemical parameters or hematological parameters between the non-contaminated haylage feeding period and contaminated haylage feeding period (data not shown).
(C) Changes in radiocesium levels in skeletal muscle, blood, and feces: The concentrations of radiocesium in the glutei skeletal muscle increased to 78–151 (113 ± 29) Bq/kg 8 weeks after the start of contaminated haylage feeding (Table 7.1). After ceasing administration of the contaminated forage, the concentration of radiocesium in the glutei skeletal muscle of horses, with the exception of one animal, decreased after 4 weeks, and 30 Bq/kg of radioactive cesium remained in the glutei skeletal muscle of this one horse. No detectable levels of radioactive cesium were noted in the glutei skeletal muscle of any horses 8 weeks after the initiation of the non-contaminated haylage administration. During the experiments, no detectable levels were found in blood samples. However, 95–136 (118 ± 14) and 234–290 (255 ± 20) Bq/kg were detected in feces samples at 4 and 8 weeks, respectively, after the start of contaminated haylage feeding. Radiocesium levels in the muscle of horses were slightly lower for horses with high levels of cesium in their feces. We suggest that the absorption capacity of radioactive cesium differs between individual horses.
(D) Changes in radiocesium levels and histopathological findings in horse organs: Two horses were sacrificed at 4, 8, and 16 weeks after the initial administration of non-contaminated haylage, and heart, spleen, liver, kidney, and psoas major muscle samples were collected for radioactive cesium measurements and histopathological evaluations. As shown in Table 7.2, no radiocesium was detected in heart, spleen, liver, kidney, or psoas major muscle samples at any time. Moreover, there were no notable histopathological findings in any horse organs (data not shown).

Table 7.1 Changes in radioactive cesium levels in horse glutei skeletal muscle obtained by biopsy, blood, and feces during the experimental period

Samples/Horse number	Before the examination	Contaminated haylage			Non-contaminated haylage			
		0 weeks	4 weeks	8 weeks	2 weeks	4 weeks	8 weeks	16 weeks
Muscle[a] (Bq/kg)								
1	–	–	–	119	–	ND	ND	ND
2	–	–	–	130	–	ND	ND	ND
3	–	–	–	151	–	30	ND	ND
4	–	–	–	122	–	ND	ND	ND
5	–	–	–	78	–	ND	ND	ND
6	–	–	–	78	–	ND	ND	ND
Blood (Bq/kg)								
1	ND[b]	ND	ND	ND	ND	ND	ND	ND
2	ND	ND	ND	ND	ND	ND	ND	ND
3	ND	ND	ND	ND	ND	ND	ND	ND
4	ND	ND	ND	ND	ND	ND	ND	ND
5	ND	ND	ND	ND	ND	ND	ND	ND
6	ND	ND	ND	ND	ND	ND	ND	ND
Feces (Bq/kg)								
1	ND[b]	ND	95	239	ND	ND	ND	ND
2	ND	ND	116	248	ND	ND	ND	ND
3	ND	ND	115	234	ND	ND	ND	ND
4	ND	ND	115	262	ND	ND	ND	ND
5	ND	ND	136	290	ND	ND	ND	ND
6	ND	ND	128	258	ND	ND	ND	ND

Muscle samples were obtained during the contaminated haylage administration period and non-contaminated haylage administration period

[a]Muscle biopsy samples were obtained at the end of the contaminated haylage administration period and 4, 8, and 16 weeks after the initiation of the non-contaminated haylage administration

[b]Not detectable: the detection limit was set to three times the standard deviation of the background

Table 7.2 Radioactive cesium levels in the heart, spleen, liver, kidney, and psoas major muscle when animals were sacrificed 4, 8, and 16 weeks after the initiation of the non-contaminated haylage administration

Horse number	Sacrifice time[a] (weeks)	Radioactive cesium level (Bq/kg)				
		Heart	Spleen	Liver	Kdney	Psoas major muscle
1	4	ND[b]	ND	ND	ND	ND
2	4	ND	ND	ND	ND	ND
3	8	ND	ND	ND	ND	ND
4	8	ND	ND	ND	ND	ND
5	16	ND	ND	ND	ND	ND
6	16	ND	ND	ND	ND	ND

[a]Horses were sacrificed 4, 8 and 16 weeks after the initiation of the non-contaminated haylage administration
[b]Not detectable: the detection limit was set to three times the standard deviation of the background

7.2.3 Discussion and Conclusion

Horses (body weight ~400 kg) were fed with radiocesium (4800 Bq/400 kg of body weight/day) for 8 weeks, and subsequently radiocesium was detected in their glutei skeletal muscles (113 ± 29 Bq/kg). Radiocesium could not be detected in these muscles 8 weeks after ceasing administration of the contaminated feed.

We examined the new tolerance level of radioactive cesium values in feed for horses (less than 100 Bq/kg). We confirmed that when horses were fed for 8 weeks with haylage contaminated with radioactive cesium at a level that was approximately 5 times higher (480 Bq/kg) than the new tolerance value, radiocesium levels in the horse meat were not high (113 ± 29 Bq/kg). The concentrations of radiocesium in the skeletal muscles of some horses were lower than the new reference/regulation level set by the Japanese Government (100 Bq/kg for radiocesium). Horses were subsequently given feed containing no radiocesium, and the concentrations of radiocesium in the muscles rapidly decreased. Concentrations of radiocesium 4 weeks after ceasing the feeding of contaminated forage decreased to background levels of less than 5 Bq/kg in 5 horses, but not in one horse.

Our results indicate that if farmers provide feed with radiocesium levels that are less than the new tolerance value, the radiocesium levels in horse meat will always be less than the new reference/regulation level. Moreover, if farmers give horses feed containing higher levels of radiocesium than the new reference/regulatory value, then they if they keep these horses on a non-contaminated diet for the same period of time that the horses were given the contaminated feed, then the radiocesium should be reduced to allowable levels in the meat.

There were two main routes of radionuclide contamination of farm animals after the nuclear accident: inhalation and ingestion of contaminated feed and/or water. Beresford and Howard (2011) reported that inhalation and water intake by animals were the most important routes in the early phase of the nuclear accident. However, at the ranch of the Animal Resource Science Center (approximately 140 km southwest from the Fukushima Daiichi nuclear power plant), the concentration of radionuclides in the air was not at a detectable levels 2 months after the accident. No radionuclide contamination was detected in drinking water. Intake via water was previously shown to be a small contributor (Nisbet et al. 2010). Therefore, in the present study the most important pathway for horses was identified as the ingestion of radionuclide-contaminated feed.

The degree of absorption from the gastrointestinal tract is an important factor that determines radiocesium levels in animal tissues. In the case of radiocesium, the source of ingested radiocesium is a major factor determining subsequent concentrations in tissues, with the true absorption coefficient ranging from 0.10 to 0.80 (Howard et al. 2001). That study showed that the absorption levels of radiocesium particles and soil binding radiocesium were lower than that of radiocesium incorporated within plants. In the present study, grass was grown for approximately 2 months after the nuclear accident, harvested, dried, and prepared for haylage. The fermented grass forage used in the present study may have contained particles of radiocesium fallout, radiocesium bound to soil, and radiocesium incorporated within the plant tissues. As described above, extremely low or undetectable levels of radiocesium were noted in blood samples, whereas increasing levels of radiocesium (118 ± 14 and 255 ± 20 Bq/kg at 4 and 8 weeks after the initial feeding of contaminated haylage, respectively) were found in feces samples. Although body weights did not change in horses given a certain amount of radiocesium in haylage during the experimental period, radiocesium levels increased in their feces. Furthermore, radiocesium levels were low in the muscles of horses with high radiocesium levels in their feces. This result indicated that the absorption capacity of radioactive cesium differed among individual horses. We previously reported differences in radiocesium metabolism among individual cows (Hashimoto et al. 2011; Manabe 2012; Manabe et al. 2011, 2013, 2014; Takahashi et al. 2012). These cows were given radiocesium-contaminated feed, and once the cows were removed from radiocesium-contaminated forage, radiocesium concentrations in their milk rapidly decreased. The rate by which radioactive nuclides was lost from milk was termed the biological half-life, which is defined as the time required for the radionuclide activity concentration in milk to be reduced by one half excluding physical decay. The uptake and loss rates of radionuclides varied among cows and tissues. The biological half-life for radiocesium has been associated with the metabolic turnover rate of cesium. Changes in radiocesium levels in milk after the Chernobyl nuclear power plant accident were summarized by the International Atomic Energy Agency (IAEA) (2005). The rate of decline in radionuclide activity concentrations in milk of different species of dairy ruminants was rapid. The half-life of radiocesium in various milk sources range 0.5–3.5 days. These values were consistent with our result (approximately 2.0 days). The biological half-life of radiocesium in animals is an important factor influencing the effectiveness and practicality of many countermeasures targeting animal-derived foodstuffs, including decontamination using “clean feeding” or cesium binders.

In order to scientifically understand the phenomenon of radiocesium in horses, more detailed research is needed on the absorption of radiocesium from feed, its accumulation and distribution in the body, its pharmacokinetics, and the mechanisms underlying its excretion from the body. In the present study, we did not examine changes in radioactive strontium-90, which accumulates in bones and plays a role as a carcinogenic factor in meat. Further research on transfer kinetics and the coefficient of strontium-90 from forage contaminated by the fallout from the Fukushima Daiichi nuclear power plant accident to horse meat is also required.

Moreover, the present results involved the initial phase of the accident; therefore, further research is needed to remediate livestock management for radiocesium contamination under the existing exposure conditions, as follows.

1. Improvements in pasture: Feasible and suitable surface improvements in grassland and radical improvements in meadow methods for Japanese farming practices should be developed to reduce radiocesium contamination in grassland.
2. Reductions in radiocesium intake: Feasible and suitable agents should be explored to prevent gut absorption by administering radiocesium binding agents to animals. The effectiveness and feasibility of adding candidate binding compounds to fodder such as clay minerals (bentonites, vermiculites, and zeolites) that adsorb cesium ions, to reduce the gut uptake of radiocesium by farm animals have been evaluated. The effectiveness and feasibility of the addition of hexacyanoferrate to feedstuffs has also been determined. The hexacyanoferrate compound, Prussian blue, is a radiocesium binder that is added to farm animal feed to reduce the transfer of radiocesium to animal products by reducing its absorption in the gut (IAEA 1997). Ammonium-hexacyanoferrate (AFCF), the commonly used form for remediation, was used extensively after the Chernobyl accident in Russia, Ukraine, and Belarus as well as in western European countries, including Norway and Sweden, and was shown to be very effective for animal use (Pearce 1994; Nisbet et al. 2010). Moreover, the acceptability of these strategies for Japanese farmers also needs to be investigated.
3. Collateral safety of animal products: "clean feeding" management has been considered feasible and suitable for the production of animal products without disrupting normal farming practices. Farm animals are provided with uncontaminated feed or forage with acceptable levels of radiocesium. To prevent the radiocesium contamination of animal products by ensuring that feedstuffs that are too highly contaminated are not ingested by farm animals, a registration system to identify individual farm animals should be developed.

In conclusion, the results of the present study demonstrate that when horses receive less than 500 Bq/kg, they produce meat contaminated with approximately 100 Bq/kg radiocesium, and that "clean feeding" can reduce radiocesium contamination levels in meat.

7.3 The Effects of "Clean Feeding" Management on Sheep Meat: Removal of Radioactive Cesium Due to the Nuclear Power Plant Accident from Mutton

One hundred sheep (50 male and 50 female), which had been kept in a paddock or livestock barn in the central area of Fukushima Prefecture (approximately 60 km from the Fukushima Daiichi nuclear power plant) for approximately 1.5 years after the nuclear accident were used in the present study (Fig. 7.1). The skeletal muscles of these animals were considered to contain higher levels of radioactive cesium than the new reference/regulation value in foods (100 Bq/kg); therefore, these animals were collected and kept in the "clean feeding" livestock barns of the National Livestock Breeding Center (Fukushima, Japan). They were given non-contaminated feed and drinking water for 3 months. Between November 19, 2012 and December 12, 2012 (10 times: 10 animals for each sampling time), animals were sacrificed under deep anesthesia to collect organ samples [skeletal muscles (quadriceps muscle and psoas major muscle), liver, kidney, spleen, genital organs (ovaries or testes), blood, and urine samples]. Radioactive cesium levels were measured in each sample using a germanium semiconductor detector, and

Fig. 7.1 Sheep kept in a paddock on a mountain in Fukushima prefecture (~60 km from the nuclear power plant) for ~1.5 years after the accident (**a**). They were kept in the "clean feeding" livestock barns of the National Livestock Breeding Center and were given non-contaminated feed (**b**). Organ samples were collected in November and December 2012 (**c** and **d**)

each nuclide was identified by gamma-ray spectrometry. No detectable levels of radiocesium were noted in skeletal muscles, liver, kidney, spleen, genital organs, ovaries, testes, blood, or urine samples.

After the sheep were directly exposed to radioactive radiocesium, they were given non-contaminated feed (so-called clean feeding). Radioactive cesium was found to have initially accumulated in sheep organs but was not detected 3 months later. These results confirm that "clean feeding" management was also effective for sheep.

7.4 The Effects of "Clean Feeding" Management for Dairy Cows: Removal of Radiocesium Due to the Nuclear Power Plant Accident from Milk

In this section, we described the effectiveness of "clean feeding" management for cows' milk. The details of this experiment have already been reported (Hashimoto et al. 2011; Manabe 2012; Manabe et al. 2011, 2013, 2014; Takahashi et al. 2012).

Changes in radiocesium levels in milk produced by cows given radiocesium contaminated feed after the nuclear accident on March 11, 2011, were examined between May 16 and June 26, 2011. Italian ryegrass, which was seeded on September 2010 and cultivated in the fields of the Animal Resource Science Center of the University of Tokyo (approximately 140 km south-west of the power plant), was harvested 2 months after the nuclear accident and prepared for fermented grass forage, haylage. The cows were born and kept at the Animal Resource Science Center and were given commercial mixed feed (total mixed ration forage: TMR) purchased from Zen-Noh Feed (Tokyo, Japan) that contained no radioactive iodine-131, cesium-134, or cesium-137 for 2 weeks before being examined. The raw materials of TMR were as follows: 45 % maize, a 29 % mixture of wheat bran and rice bran, a 21 % mixture of soybean oil cake and rapeseed oil cake, and 5 % additives (minerals and vitamins). The feed ingredients of TMR were as follows: 16 % crude protein, 2.5 % crude fat, 10 % crude fibers, 10 % crude ash, 0.8 % calcium, and 0.5 % phosphorus. The cows were initially given the mixed feed of haylage and TMR at 10 and 25 kg/600 kg of body weight/day, respectively for 2 weeks, and then only TMR (35 kg/600 kg of body weight/day) for 4 weeks as "clean feeding" management. During this experiment, milk was collected twice a day and mixed for each cow. The weight and radioactive cesium levels for the mixed milk from each cow were measured daily. No radioactive iodine-131 was detected in either the milk or haylage. The radiocesium level in the mixed feed of haylage and TMR was 380 Bq/kg. Four days after the initial administration of contaminated feed, radiocesium concentrations in the milk rapidly increased to 30 Bq/kg, and then equilibrated to 36 Bq/kg after 12 days (lower than the new reference/regulation value for milk: 50 Bq/kg) (Table 7.3). Cows were then given TMR containing no radiocesium ("clean feeding"), and the concentrations of

Table 7.3 Changes in radiocesium concentrations in milk

Days after initiation	Radioactive cesium level in milk (Bq/kg)	
	Control group	Treated group
	Non-contaminated feed	Contaminated feed
0	ND[a]	ND
2	ND	22 ± 4[b]
4	ND	29 ± 5
6	ND	33 ± 5
8	ND	34 ± 4
10	ND	35 ± 6
12	ND	36 ± 4
14	ND	36 ± 5
	Non-contaminated feed	Non-contaminated feed
16	ND	29 ± 3
18	ND	15 ± 3
20	ND	12 ± 4
22	ND	9 ± 2
24	ND	8 ± 3
26	ND	6 ± 2
28	ND	ND
42	ND	ND
56	ND	ND

[a]Not detectable: the detection limit was set to three times the standard deviation of the background
[b]Each value represents the mean ± standard deviation (n = 6)

radiocesium in the milk rapidly decreased, with undetectable levels (less than 5 Bq/kg: background level) noted 14 days after the initial administration of the non-contaminated feed. Concentrations of radiocesium in milk were then maintained at the background level. When cows (body weights ~ 600 kg) were given contaminated feed (12,600 Bq/600 kg body weight/day), 5.71 % of radiocesium was secreted into the milk (720 Bq/20 kg milk/day). The transfer coefficient (Fm) for the transfer of radiocesium from cow feed to milk was calculated according to the following formula: Fm (day/L) = radiocesium level in milk (Bq/L)/the intake of radiocesium by each cow (Bq/cow/day) (Hashimoto et al. 2011; Manabe et al. 2013, 2014; Takahashi et al. 2012). The radiocesium Fm value was 0.0029 (day/L) at the highest point of radiocesium levels in the milk.

Generally, infants and schoolchildren consume milk daily; therefore, advanced safety measures are necessary to ensure that milk contains extremely low levels of radiocesium. It is thus essential that cows do not orally ingest feed contaminated with radioactive cesium. Further research is needed to ensure the safety of domestically produced milk. The dynamics of more than 90 % of the radiocesium contained in polluted feed that is not secreted into the milk needs to be elucidated in more detail. It is currently unknown whether radioactive cesium that is taken into the cow body is rapidly excreted through the feces, urine, sweat, or bile. Nor do we know where, or in which organs, or how much radiocesium accumulates in the cow body if it is not rapidly excreted. We do not know whether the radioactive cesium that accumulates in a cow's body is transferred to milk. Further studies are needed to understand the molecular mechanisms regulating cesium secretion into milk in dairy cows. The concentrations of various components including minerals in milk are strictly maintained within a constant range in mammals, and cesium is also strictly controlled. The concentrations of potassium, an essential element, are maintained at approximately 1.5 mg/g in cow's milk. The dynamics of cesium in organisms is considered to be similar to those of potassium. If the secretion dynamics of cesium into milk are similar to those of potassium, radioactive cesium levels in milk may also be maintained at a certain level. Further research is needed not only for the reconstruction of livestock industries, which were debilitated by the nuclear power plant accident, but also for the protection and improvement of the health of the Japanese people.

In conclusion, dairy cows given radiocesium-contaminated feed produced milk that was contaminated with radiocesium. When these cows were given non-contaminated feed (so-called clean feeding), radioactive cesium levels in milk rapidly decreased and reached undetectable levels after 4 weeks. These results confirm that "clean feeding" management is also effective for dairy cows.

7.5 Conclusion

In the present studies, we demonstrated the effects of "clean feeding" management on livestock products that were contaminated with radioactive cesium due to the Fukushima Daiichi nuclear power plant accident in March 2011. Radiocesium-contamination in animal products represents an important potential route of exposure in the human food chain; therefore, the development solutions reduce and combat radiocesium-contamination is a serious social issue in Japan. Most farm animals are kept in closed barns in Japan, which reduced the initial contamination of animal products by radioactive nuclides in the early phase of the nuclear power plant accident. Moreover, pigs and chickens in Japan were given imported feed that was not contaminated by radioactive nuclides. However, adequate amounts of grass feed, at least 30 % or more of the total feed, are essential for rearing herbivorous animals (such as horses, sheep, and cattle), and uncontaminated grass feed and/or forage containing acceptable amounts of radiocesium are needed. The present results concerning products from horses, sheep, and cattle revealed that no detectable level of radiocesium was noted in the products (meat or milk) of herbivores that received radiocesium-contaminated feed, followed by non-contaminated feed, a livestock management system called "clean feeding". In conclusion, "clean feeding" management plays a crucial role in solving this social problem.

References

Beresford NA, Howard BJ (2011) An overview of the transfer of radionuclides to farm animals and potential countermeasures of relevance to Fukushima releases. Integr Environ Assess Manag 7:382–384

Hashimoto K, Tanoi K, Sakurai K, Hashimoto T, Nogawa N, Higaki S, Kosaka N, Takahashi T, Enomoto Y, Onoyama I, Li JY, Manabe N, Nakanishi T (2011) The radioactivity measurement of milk from the cow supplied with pasture grown in Ibaraki-prefecture, after the Fukushima nuclear power plant accident. Radioisotopes 60:335–338

Howard BJ, Beresford NA, Voigt G (2001) Countermeasures for animal products: a review of effectiveness and potential usefulness after an accident. J Environ Radioact 56:115–137

International Atomic Energy Agency (IAEA) (1997) The use of Prussian blue to reduce radiocesium contamination of milk and meat produced on territories affected by the chernobyl accident, IAEA=TECDOC-926. IAAEA, Vienna

International Atomic Energy Agency (IAEA) (2005) Environmental consequences of the chernobyl accident and their remediation: twenty years of experience report of the UN chernobyl forum Expert Group "Environment" (EGE). IAEA, Vienna

Manabe N (2012) Metabolism of radioactive cesium in dairy cows. Chem Biol 50:668–670

Manabe N, Li JY, Takahashi T, Endo M, Enomoto Y, Tanoi K, Sakurai K, Nakanishi T (2011) Transition to the milk of radioactive materials in feed for dairy cattle and future pollution control. Dairy Jpn 12:25–27

Manabe N, Takahashi T, Li JY, Tanoi KM, Nakanishi T (2013) Changes in the transfer of fallout radiocesium from pasture harvested in Ibaraki prefecture, Japan, to cow milk two months after the Fukushima Daiichi nuclear power plant accident. Springer, Germany

Manabe N, Takahashi T, Li JY, Tanaka T, Tanoi K, Nakanishi T (2014) Farm animals and livestock products contamination with radioactive cesium due to the Fukushima Daiichi nuclear power plant accident. Sustain Livest Prod Hum Welf 68:1085–1090

Nisbet AF, Brown J, Howard BJ, Beresford NA, Ollagnon H, Turcanu C, Camps J, Andersson K, Rantavaara A, Lkaheimonen T, Duranova T, Oughton D, Kirchner G, Papachristodoulou C, Ioannides K, Kwakman P (2010) Decision aiding handbooks for managing contaminated food production systems, drinking water and inhabited areas in Europe. Radioprotection 45:S23–37

Pearce J (1994) Studies on any toxicological effects of Prussian blue compounds in mammals: a review. Food Chem Toxicol 32:577–582

Takahashi T, Enomoto Y, Endo M, Onoyama I, Tomimatsu S, Ikeda M, Li JY, Tanoi K, Nakanishi T, Manabe N (2012) Changes in radionuclide levels in milk from the cow supplied with pasture grown in Ibaraki prefecture, after the accident in Fukushima Daiichi nuclear power plant. Radioisotopes 61:551–554

Chapter 8
Adverse Effects of Radiocesium on the Promotion of Sustainable Circular Agriculture Including Livestock Due to the Fukushima Daiichi Nuclear Power Plant Accident

Noboru Manabe, Tomotsugu Takahashi, Chunxiang Piao, Junyou Li, Keitaro Tanoi, and Tomoko M. Nakanishi

Abstract Appropriate treatment is urgently needed for farm and ranch waste that has been contaminated with radioactive cesium from the Fukushima Daiichi nuclear power plant accident. We previously developed an aerobic ultra-high temperature fermentation (more than 115 °C) method to inhibit intestinal infectious diseases. Fermented waste (compost), in which pathogens were sterilized, was useful as a fertilizer to grow crops. In the present study, we examined the kinetics of radioactive cesium in farming fields to promote sustainable circular agriculture including livestock in farms and pasture fields in wide areas of Tohoku and Kanto, in an approximately 200 km radius from the nuclear power plant. The compost produced at the experimental ranch of the Animal Resource Science Center, the University of Tokyo, was contaminated with radioactive cesium (approximately 900 Bq/kg). Some crops (soybean, sweet corn, eggplant, bitter gourd, potato, cabbage, and ginger) were cultivated in cubic holes (approximately $1 \times 1 \times 1$ m) that were filled with contaminated compost in the field of the experimental ranch. Each crop was planted in a hole and cultivated in an appropriate manner for a suitable time period. Radiocesium levels in the roots, stems, leaves, and fruits of each crop at harvest were lower than 20 Bq/kg, which was below the new reference/

N. Manabe (✉)
Animal Resource Science Center, Graduate School of Agricultural and Life Sciences, The University of Tokyo, Kasama 319-0206, Japan

Osaka International University, Osaka 570-8555, Japan
e-mail: n-manabe@oiu.jp

T. Takahashi • C. Piao • J. Li
Animal Resource Science Center, Graduate School of Agricultural and Life Sciences, The University of Tokyo, Kasama 319-0206, Japan

K. Tanoi • T.M. Nakanishi
Graduate School of Agricultural and Life Sciences, The University of Tokyo, 1-1-1 Yayoi, Bunkyo-ku, Tokyo 113-8657, Japan

T.M. Nakanishi, K. Tanoi (eds.), *Agricultural Implications of the Fukushima Nuclear Accident*, DOI 10.1007/978-4-431-55828-6_8

regulation value (100 Bq/kg) for food. In conclusion, when crops were planted using compost contaminated with radioactive cesium (900 Bq/kg; more than twice the new tolerance value of 400 Bq/kg for fertilizer and compost), the radiocesium levels in the crops were as low as one-fifth of the new reference/regulation value.

Keywords Aerobic ultra-high temperature fermentation • Compost • Crop contamination • Fertilizer • Livestock • Manure processing • Sustainable circular agriculture • Radiocesium • The Fukushima Daiichi nuclear power plant accident

8.1 Introduction

The greatest invention of the twentieth century is considered to be the ability to produce ammonia industrially in large quantities at a low cost. Ammonia was produced from nitrogen in the air by the Haber–Bosch process in large factories. The nitrogen-based chemical fertilizers, ammonium sulfate, ammonium nitrate, and urea, were stably and inexpensively provided, and they markedly increased food production. If this ammonia synthesis method had not been invented, the population of the world would have remained at less than a quarter of the current population. However, the provision of a sufficient amount of organic fertilizers, in addition to chemical fertilizers, is known to be essential for the continuous and stable production of crops. Thus, compost is an important agricultural material for promoting sustainable circular agriculture including livestock.

A devastating earthquake occurred in the Kanto and Tohoku regions, northeastern areas of Japan, on March 11, 2011, which also led to the Fukushima Dai-ichi nuclear power plant accident. The Japanese government subsequently took a series of measures to deal with the consequences of this accident. New reference/regulation values were established by the Ministry of Health, Labour, and Welfare, Japan (MHLW) on April 1, 2012. Radioactive cesium will hereafter be used to refer to radionuclides, and radiocesium levels include the total values for ^{137}Cs and ^{134}Cs. Radiocesium levels in drinking water, in milk and infant foods, and in other foods (including cereals, vegetables, fruits, meat, eggs, and fish) had to be maintained at less than 10, 50, and 100 Bq/kg, respectively. These new reference/regulation values remain the standard today. In response to these new reference/regulation values, the Ministry of Agriculture, Forestry and Fisheries, Japan (MAFF) published new tolerance values on March 23, 2012 in which they limited the acceptable radioactive cesium levels in livestock feed and bedding, and in fertilizers such as compost for crop production (MAFF 2012). Radioactive cesium levels in feed for meat production had to be maintained at less than 100 Bq/kg for dairy cattle and horses, 80 Bq/kg for pig feed, 160 Bq/kg for chicken feed, and 40 Bq/kg for farmed fish. In addition, radiocesium levels within livestock bedding, fertilizers and compost, soil for crop cultivation, and soil-improving materials had to be maintained at less than 400 Bq/kg. Because Japanese people have characteristic eating habit and farming systems, fundamental research to reflect these Japanese styles and animal feeding methods is needed for revisions of reference/regulation values and tolerance values.

Radioactive cesium crop absorption is considered to compete with potassium absorption; thus, an excessive potassium fertilization technique has been developed to reduce the levels of radiocesium absorption. Appropriate amounts of phosphorus- and nitrogen-based chemical fertilizers were used in this technique, but no organic fertilizers (such as compost produced from the feces of livestock). If farmers continue to use only chemical fertilizers for a long period of time, then the decreases occur in crop production. An appropriate amount of an organic fertilizer is therefore essential for stable agricultural production.

We determined the quantities of radioactive cesium (released by the nuclear power plant accident) that contaminated plants for food and feed and quantities that migrated from feed to livestock and livestock products. We also determined the amount of radiocesium-contaminated feces produced by livestock, the amounts of radiocesium that moved to soil from fermented feces (compost), and the amount that was transferred to plants from soil fertilized with compost. In addition, we summarized the adverse effects of radiocesium on the promotion of sustainable circular agriculture including livestock.

8.2 Aerobic Ultra-High Temperature Fermentation Technique for Livestock Feces

Over the past 10 years, we have developed a novel, aerobic, ultra-high temperature fermentation technique for livestock feces at the experimental ranch of the Animal Resource Center, The University of Tokyo (Kasama, Japan) to prevent the spread of digestive tract infectious diseases (Manabe et al. 2014a, b). Briefly, we built 4 fermenters (4.3 m width × 2.4 m height × 8 m depth) surrounded on 3 sides by a wall and a floor made of concrete. Two grooves (10 cm × 10 cm) were dug into the floor. A vinyl chloride tube (9 cm in diameter) for delivering air was placed inside each groove. Air was continuously sent through this tube from the blower, which was placed outside the fermenter. The fermentation process was as follows (Fig. 8.1). Fermentation feedstock (including the excretions of farm animals, litter residues, residues of livestock feed, and residues of crop silage and haylage) was mixed with an equal amount of the end products of fermentation. These farm residues were contaminated with radioactive cesium due to the nuclear power plant accident. An appropriate amount of air was supplied during the fermentation process. Three days after initiating fermentation, the temperature of the central area of the fermenter increased to more than 115 °C. Denaturing gradient gel electrophoresis analysis revealed that the dominant bacteria in the flora of the ultra-high-temperature fermenter were the *Geobacillus* family (*Geobacillus thermodenitrificans*, *Geobacillus tropicalis*, and *Geobacillus stearothermophilus*), *Bacillus thermodenitrificans*, *Sphingobacteriaceae bacterium*, *Thermoactinomyces*

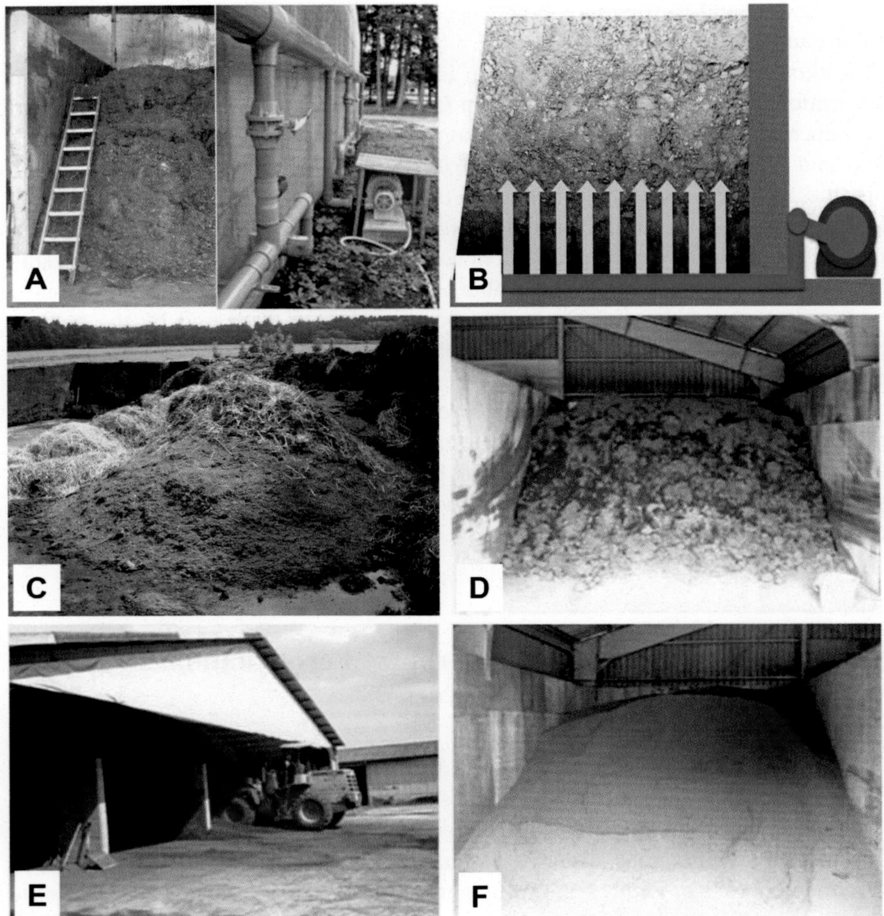

Fig. 8.1 Aerobic ultra-high temperature fermentation process for livestock feces and farm residues. (**a** and **b**) Overview of fermenters (4.3 m width × 2.4 m height × 8 m depth). Three side walls and a floor were made of concrete. An appropriate amount of air was supplied from 2 vinyl chloride tubes installed in the floor grooves of the floor without interruption from the electric blower, which was placed outside of the fermenter. (**c**) Feedstock (mixture of the excretions of farm animals, litter residues, livestock feed residues, and residues of crop silage and haylage) before fermentation. (**d**) Appearance of the fermenter of the start of the fermentation stage. Feedstock was mixed with an equal amount of the end products of fermentation (mostly the dormant spores and sprouts of high temperature fermentation bacteria). (**e**) During fermentation, all fermenting mixtures were moved into a neighboring fermenter using a wheel loader. This mixture process of fermenting substances was repeated every 7 days. (**f**) After being mixed 6 times, the fermenting substances became powdery (moisture content less than 35 %) 7 weeks after the start of fermentation. The fermented product at the final point was used as an organic fertilizer for crops and pasture cultivation

sanguinis, *Thermus thermophilus*, *Thermaerobacter composti*, and *Bacteroidetes* bactereria. Bacteria from the *Planifilum*, *Rhodothermus*, *Thermaerobacter*, and *Thermus* families were also detected. The ultra-high temperature continued for 2 or 3 days. Seven days after the initiation of fermentation, all of the fermented mixture was moved into the neighboring fermenter using a wheel loader. This exchange was repeated six times. The fermented product at the final point (7 weeks after initiation) became powdery (moisture content less than 35 %). Most of the end products of the final fermentation were mixed with the next fermentation feedstock, and the next fermentation process was started. The end products were used as an organic fertilizer for crops and pasture cultivation at the experimental ranch. The radioactive cesium contamination level in the compost was approximately 900 Bq/kg, more than twice the new tolerance value (400 Bq/kg). Pathogens were sterilized in the fermented waste, which inhibited the spread of intestinal infectious diseases. In our preliminary experiments, aerobic ultra-high temperature-fermented compost exhibited excellent effects as an organic fertilizer on 22 different crops including rice, soybean, kidney bean, green peace, onion, green onion, cucumber, tomato, eggplant, potato, pepper, cabbage, radish, and turnip.

8.3 Radiocesium Dynamics in Crops Grown with Contaminated Compost

Prior to the nuclear accident, the average level of radioactive cesium in farmland soil in Japan was approximately 20 Bq/kg (between 5 and 140 Bq/kg). After the accident, the Japanese government established a new tolerance value (400 Bq/kg) for farmland soil, soil improvement materials, and fertilizers. An appropriate method for treating farm waste contaminated with radioactive cesium has not yet been developed. In order to obtain fundamental understanding for the development of such a method, we examined radiocesium dynamics in crops. These crops were cultivated using aerobic ultra-high temperature-fermented compost produced at the ranch of the University of Tokyo, which is located approximately 130 km southwest of the Fukushima Daiichi nuclear power plant.

Approximately 3 months after the accident, June 13, 2011, ^{131}I was not detected in the soil of farm fields, although radiocesium levels in the soil were 50–240 Bq/kg. We prepared haylage of Italian ryegrass, which was grown in farm fields. The radioactive cesium level in the haylage was 3900 Bq/kg (the new tolerance value for cattle, goat, sheep, and horse feed was less than 100 Bq/kg). Our preliminary experiment performed in June, 2011 showed that radioactive cesium levels in the skeletal muscle of goats (native Japanese Shiba goat: body weight of approximately 40 kg) were approximately 130 Bq/kg (the new reference/regulation value for humans was 100 Bq/kg) when they were fed this contaminated haylage (60 g/kg of body weight/day: 2.4 kg/goat/day, and 9360 Bq/goat/day) for 1 month (Manabe et al. 2014c). Radioactive cesium levels in the feces of these goats (approximately 1 kg/goat/day) were approximately 10,000 Bq/kg.

Between June and September, 2011, approximately 40 milking cows, 120 goats, 20 horses, and 30 pigs were reared in the experimental ranch of the University of Tokyo. The feces, straw litter residues, feed residues, and the waste feed of livestock were used to produce the fermented compost. Radioactive cesium levels in aerobic ultra-high temperature-fermented compost were approximately 900 Bq/kg.

We examined the circulation of radioactive cesium in crops, using this compost as a crop fertilizer, by the following process (Fig. 8.2). First, we dug holes (approximately 1 m × 1 m × 1 m) in a field at the experimental ranch. Second, each hole was filled with compost contaminated with radiocesium (900 Bq/kg). Third, each crop (soybean, sweet corn, eggplant, bitter gourd, potato, cabbage and ginger) was planted in a cubic hole and cultivated in an appropriate manner for a sufficient period of time. Finally, we measured radiocesium levels in the roots, stems, leaves and fruits of each crop were at each crop's time harvest.

Radioactive element concentrations in each sample were measured using a germanium semiconductor detector, and each nuclide was identified by gamma-ray spectrometry. ^{134}Cs and ^{137}Cs were quantified at 604.7 and 661.6 keV, respectively, and each Bq value was then calculated by the calibration of count values. Each radionuclide concentration was calculated based on the weight of each sample. The detection limit was set to three times the standard deviation of the background. The radioactive cesium levels included both ^{134}Cs and ^{137}Cs levels.

Radioactive cesium levels in the roots, stems, leaves, or fruit of each crop were lower than 20 Bq/kg (1/5 of the new reference/regulation value for human food). Thus, the present empirical research results confirmed that radiocesium levels in roots, stems, leaves, or fruit of crops cultivated in radiocesium-contaminated compost were less than one fifth of the reference/regulation value. These results suggested that the transition rate of radioactive cesium to crops from soil and compost was low.

We also conducted preliminary experiments. Briefly, we planted and cultivated barley, buckwheat, sweet corn, dent corn, Italian rye grass, soybean, green peas, eggplant, tomato, bitter gourd, potato, sweet potato, cabbage, green onion, onion, ginger, bitter gourd, cucumber, and lotus root in soil contaminated with approximately 200 Bq/kg of radiocesium (half of the new tolerance value) mixed and fertilized with radiocesium-contaminated compost in a field at the Animal Resource Science Center. Similar to the hole-culture experiment described above, each crop was planted in the ridge of the field and was cultivated in an appropriate manner for a sufficient period of time. Radiocesium levels in the roots, stems, leaves, and fruit of each crop were measured at the harvest time of each crop. Radioactive cesium was below the detection limit in all parts of the crops examined.

The results of the present study demonstrate a low migration rate of radioactive cesium to plants from soil containing contaminated compost. However, the mechanism of radioactive cesium migration into crops has not yet been elucidated in detail. We speculate that one of the mechanisms involves the dormant spores and

Fig. 8.2 Overview of the crop cultivation experiment. (**a** and **b**) One cubic meter holes were dug and filled with radiocesium-contaminated compost (approximately 900 Bq/kg). (**c** and **d**) Cultivation of soybean and sweet corn, respectively, in each cubic hole. (**e** and **f**) Eggplant was planted in the cubic hole and was cultivated in an appropriate manner for a sufficient period of time. Radiocesium levels in the roots, stems, leaves, and fruit of each crop were measured at the harvest time

sprouts of high temperature fermentation bacteria. During fermentation, bacteria incorporate radioactive cesium into their bodies. Radioactive cesium then strongly binds with fine and stable organic and inorganic materials in the bacterial body. Since the radioactive cesium remained strongly and stably bound to bacterial materials after compost was mixed with the soil, radiocesium could not be easily absorbed into the plant.

8.4 Conclusion

Crops were grown in soil contaminated with radioactive cesium (900 Bq/kg; twice the new tolerance value for soil). The radioactive cesium levels in the roots, stems, leaves, or fruits of each crop were less than 20 Bq/kg, which is lower than the new reference/regulation value for human food, 100 Bq/kg. Moreover, when a range of crops was cultivated in soil contaminated with radioactive cesium (approximately 200 Bq/kg, i.e. half of the new tolerance value for soil), radioactive cesium was not detected in the roots, stems, leaves, or fruits of each crop. These practical research results indicate that the tolerance values of radiocesium for agriculture should be reviewed. Sustainable agriculture, including livestock, need to be revived within radiocesium-contaminated areas of Japan, as they are important food production regions. The utilization and production of livestock compost play an important role in such sustainable cycling agriculture system. In conclusion, the results of the present study contribute to the revival of cycling agriculture.

References

Manabe N, Takahashi T, Li JY (2014a) Future of livestock waste processing with an emphasis on safety of livestock products. Sustain Livest Prod Hum Welf 68:447–451

Manabe N, Takahashi T, Li JY, Tanaka T, Tanoi K, Nakanishi T (2014b) Effect of radionuclides due to the accident of the Fukushima Daiichi nuclear power plant on the recycling agriculture including livestock. New Food Ind 56:45–50

Manabe N, Takahashi T, Li JY, Tanaka T, Tanoi K, Nakanishi T (2014c) Farm animals and livestock products contamination with radioactive cesium due to the Fukushima Daiichi nuclear power plant accident. Sustain Livest Prod Hum Welf 68:1085–1090

Ministry of Agriculture, Forestry and Fisheries of Japan (MAFF) (2012) Setting the tolerance improvement materials and feed fertilizer including radioactive cesium: 2. Allowable radioactive cesium in the feed. http://www.maff.go.jp/j/syouan/soumu/saigai/supply.html (in Japanese)

Chapter 9
Wild Boars in Fukushima After the Nuclear Power Plant Accident: Distribution of Radiocesium

Keitaro Tanoi

Abstract In the present chapter, I present the distribution of radiocesium in wild boars as well as the official monitoring data of wild boars from Fukushima. After the nuclear accident in 2011, the radiocesium contamination levels in wild boars from most places in Fukushima Prefecture exceeded 100 Bq/kg. The most contaminated wild boars were observed in Soso district where the radiocesium concentration in the soil was the highest in the entire Fukushima Prefecture. To understand radiocesium contamination in wild boars in more detail, we measured radiocesium concentrations in different organs and tissues of wild boars inhabiting Iitate Village in Soso district in 2012 and 2013. After capturing the wild boars, we collected 24 organs and tissues and put them into vials. Using an NaI(Tl) scintillation counter, we determined the concentrations of radiocesium (^{134}Cs and ^{137}Cs) and found that the levels were highest in the muscles (approximately 15,000 Bq/kg) and lowest in the ovaries (approximately 600 Bq/kg) in 2012, indicating a large variation between the organs and tissues. The trends were similar in 2012 and 2013. Observations of the contamination levels in wild boars could reveal the radiocesium availability in the forest and village ecosystem.

Keywords Fukushima Daiichi nuclear power plant accident • Wild boar • Radiocesium • Fission products • Food security • Wild animals • Wild life • Forest • Radioactive contamination

9.1 Introduction

Significant quantities of radiocesium were released during the accident at the Fukushima Dai-ichi Nuclear Power Plant (FDNPP) of the Tokyo Electric Power Company in March 2011 (Yasunari et al. 2011; Zheng et al. 2014). From soil

K. Tanoi (✉)
Graduate School of Agricultural and Life Sciences, The University of Tokyo, 1-1-1 Yayoi, Bunkyo-ku, Tokyo 113-8657, Japan
e-mail: uktanoi@mail.ecc.u-tokyo.ac.jp

T.M. Nakanishi, K. Tanoi (eds.), *Agricultural Implications of the Fukushima Nuclear Accident*, DOI 10.1007/978-4-431-55828-6_9

measurements, it was shown that radiocesium was highly deposited in the north-western region of the power plant (Saito et al. 2015). Fortunately, very small quantities of radiostrontium (Steinhauser et al. 2013) and plutonium (Zheng et al. 2012; Schneider et al. 2013) were deposited in the soil in most areas of Fukushima. Several months after the accident, the dominant radionuclides in the soil were ^{134}Cs and ^{137}Cs. There have been many kinds of inspections of agricultural products and foods in Fukushima Prefecture, and the contamination levels in the wild animals of the forests, particularly wild boars, were found to be very high, presumably because wild boars have a habit of eating soil attached to their food, similar to an earthworm. Firstly, I summarized the inspection data of wild boars from Fukushima prefecture in the next section.

Data for radiocesium distribution in animals were limited. Green et al. (1961) reported the distribution of radiocesium and potassium in pigs and calves. Fukuda et al. (2013) reported radiocesium distribution in cattle. These reports showed that radiocesium levels were the highest in muscles and that there were large variations between the organs. Here, I present the radiocesium distribution within wild boars captured in Iitate Village where the habitants remain evacuated because of the high radiation levels caused by radiocesium deposition (Fig. 9.1).

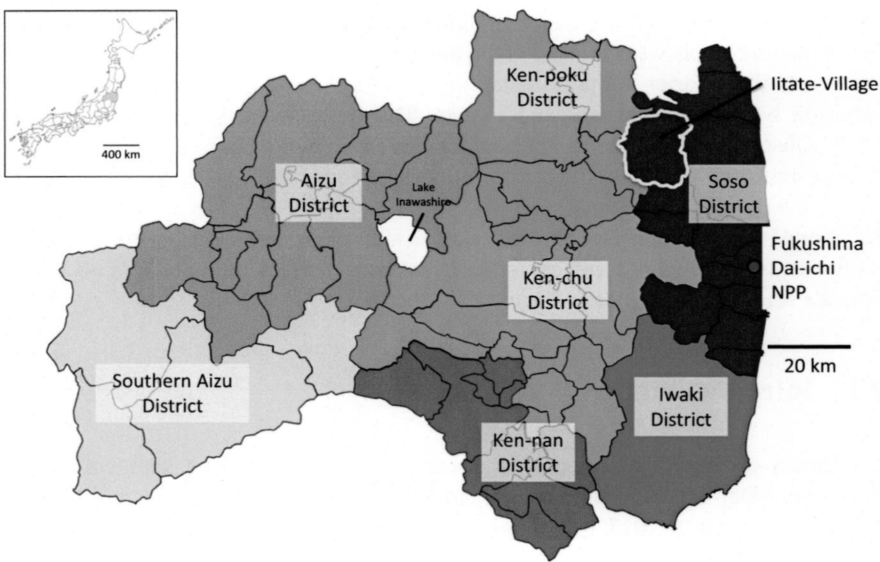

Fig. 9.1 Map of Iitate Village. Iitate Village is located in Soso district. The total area is 23,013 ha. Approximately 75 % of the area is mountain or forest

9.2 Inspection Data from Fukushima

The inspection data is accessible from the website "Fukushima Shinhatsubai" (http://www.new-fukushima.jp/monitoring/en/). I have summarized the data and separated it according to each district (Fig. 9.2). The radiocesium levels were high in Soso district, and the highest concentration of ^{137}Cs exceeded 50,000 Bq/kg in 2013. Because the radiocesium levels remained high during these 4 years, we cannot speculate about when the meat of wild boars from this area can be used as food. From April 2012, the standard in Japan for the allowable radiocesium concentration in meat has been 100 Bq/kg (Hamada et al. 2012). The current radiocesium levels in wild boar meat are now significantly higher than the standard level.

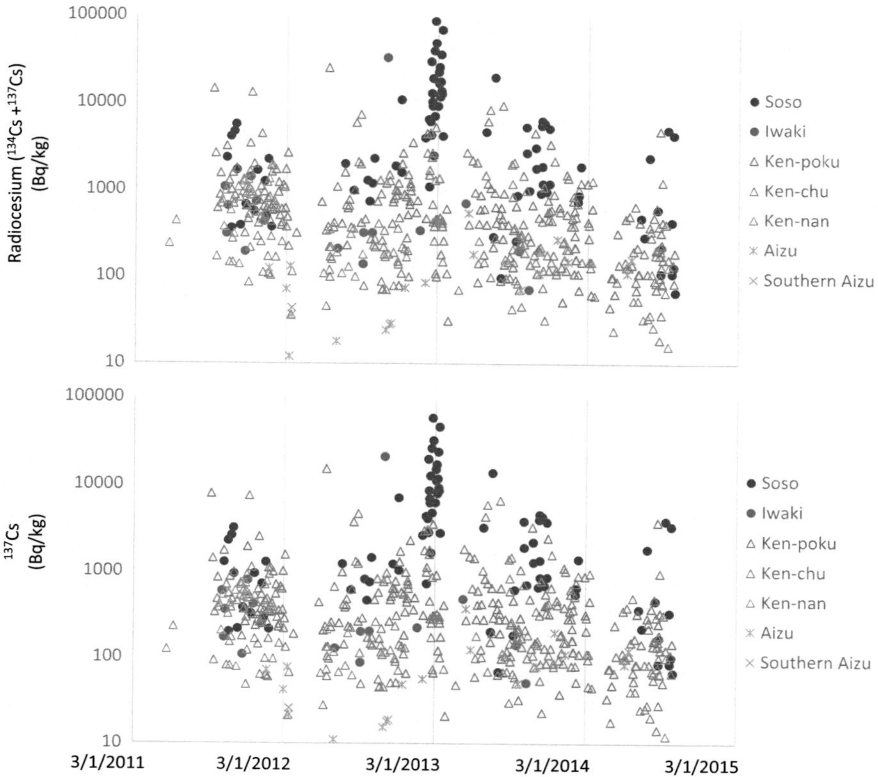

Fig. 9.2 Radiocesium concentrations in wild boars collected in Fukushima after the nuclear accident. *Upper*: Total radiocesium concentrations ($^{134}Cs + ^{137}Cs$) in wild boars, which is the measure used to monitor food in Japan. *Lower*: ^{137}Cs concentrations in wild boars

9.3 Distribution of Radiocesium in Wild Boars in 2012 and 2013

Because Iitate Village is now in the evacuation zone, damage to agricultural fields by wild animals such as monkeys and wild boars has been increasing. Before the nuclear power plant accident, hunting was performed to control the number of wild animals, and some of the hunted animals such as wild boars and deer were sold for meat. However, the radiocesium contamination in wild animals now prevents people from hunting them. The wild animal problem has become more serious with time.

In 2012, members of the NPO “Resurrection of Fukushima” and researchers from the University of Tokyo started to investigate the radiocesium levels in wild boars in Iitate Village (Tanoi et al. 2016). Wild boars were hunted for pest control and then the organs, tissues, contents of digestive organs, and blood were collected (Figs. 9.3 and 9.4). We investigated seven wild boars in 2012 and two wild boars in 2013. Five were captured in a single cage on November 25, 2012 (identified as 20121125-01 to 20121125-05) and two on November 29, 2012 (20121129-01 and 20121129-02). Two more boars were captured in separate cages on December 6, 2013 (20131206-01 and 20131206-02). Radiocesium in the samples were measured using a NaI(Tl) scintillation counter (2480 WIZARD^2 gamma counter, PerkinElmer, Waltham, MA). Only the radiocesium activities of the blood samples taken in 2012 were measured using a germanium semiconductor detector (GEM-type, ORTEC, SEIKO EG&G, Tokyo, Japan).

Wild animals hunted as vermin were consumed previously as game meat. In Japan, the radiocesium concentration in food has been controlled according to the sum of ^{134}Cs and ^{134}Cs activities (Bq kg^{-1}). In general, we eat the muscles of wild animals. The average radiocesium concentration in muscle tissues was approximately 15,000 Bq kg^{-1}, which was the highest value among the organs tested (Fig. 9.5). In addition to the muscle, most of the organs exceeded the provisional regulation value for meat (500 Bq kg^{-1}) (Hamada et al. 2012), which was the regulation level of radiocesium until March 31, 2012, although the new standard from April 1, 2012 remains at 100 Bq kg^{-1} (Hamada et al. 2012). The radiocesium concentrations in the ovaries were the lowest among the organs (600 Bq kg^{-1}); however, they still exceeded the provisional regulation value (Fig. 9.5). None of the organs were distributed as food.

The ^{137}Cs accumulation patterns were almost the same between 2012 and 2013. The highest ^{137}Cs concentrations were found in the muscles, kidneys, tongue, and heart (Figs. 9.5 and 9.6). The tissues with low ^{137}Cs were similar between 2012 and 2013. The lowest ^{137}Cs concentrations were found in the ovaries, bone, and thyroid glands. The trend is similar to that observed in cattle contaminated by the FDNPP accident in 2011 (Fukuda et al. 2013), whereby the radiocesium concentrations in the muscle, kidney, tongue, and heart were consistently higher than the other organs. In addition to cattle, the radiocesium concentrations were highest in the muscles of pigs fed with brown rice that was contaminated by the FDNPP accident,

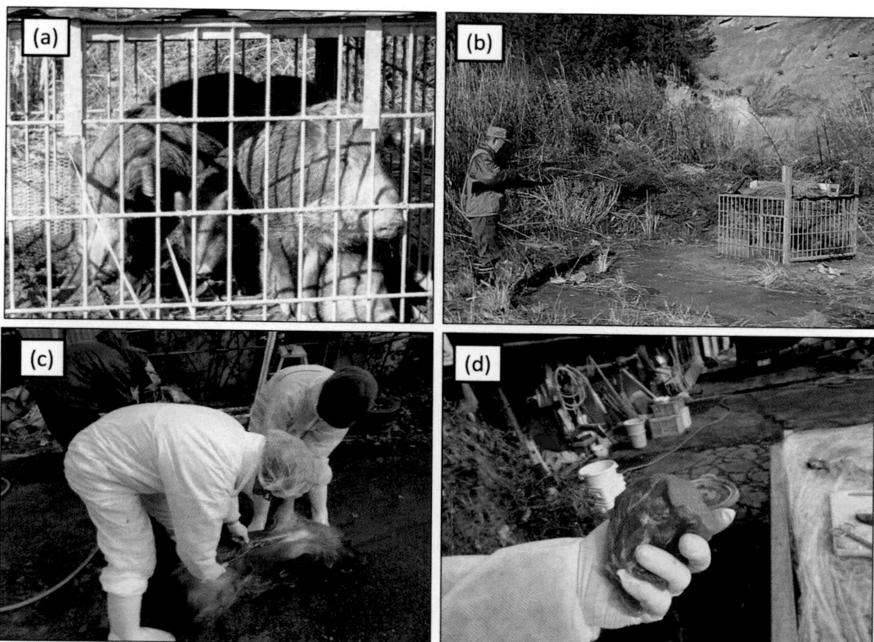

Fig. 9.3 The procedure for collecting wild boars. After capturing the boars in a cage, they were shot and dissected into organs and tissues

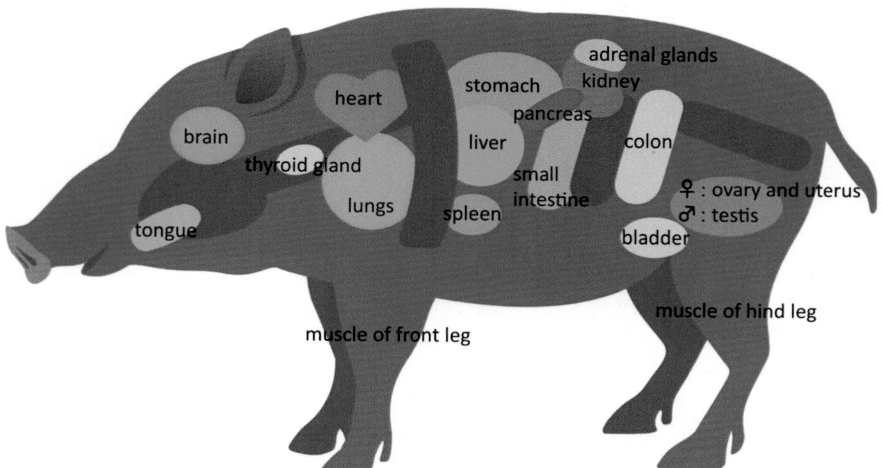

Fig. 9.4 Wild boar anatomy chart

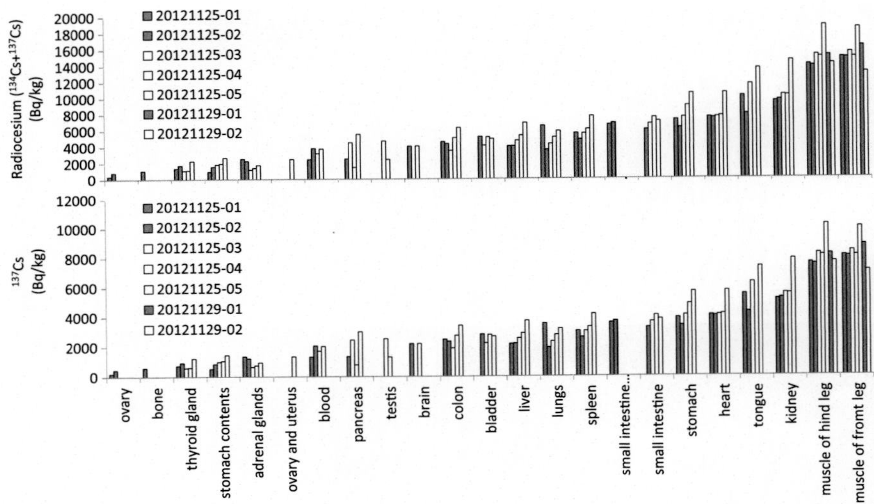

Fig. 9.5 Radiocesium concentrations in the organs and tissues of wild boars in November 2012. The *gray bars* represent adult boars and *white bars* represent juvenile boars

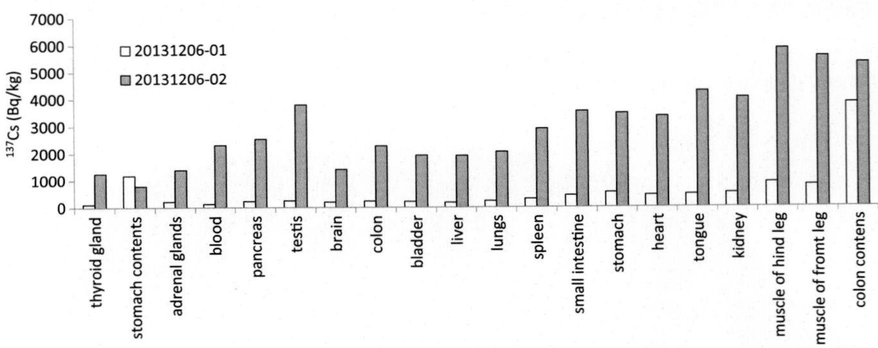

Fig. 9.6 ^{137}Cs concentrations in the organs and tissues of wild boars in winter 2013. The *gray bars* represent adult boars and *white bars* represent juvenile boars

followed by the liver and digestive tract, which follow a similar ranking to our data (Ohmori et al. 2014). In addition, when pigs were contaminated by the global fallout around 1950–1960, the ranking of contaminated organs was similar to the present study: higher concentrations in the muscle, kidney, and heart and lower concentrations in the brain, blood, and female reproductive tract (ovary and uterus in our study) (Green et al. 1961).

Wild boars eat a variety of foods such as plants, insects, mushrooms, and small animals; therefore, the radiocesium concentration in the stomach contents reflect the available radiocesium in the forest ecosystem. In the present study, the ^{137}Cs concentrations of boar stomach contents in 2012 and 2013 were almost the same

Table 9.1 The ratios of blood to muscle in ^{137}Cs

Animals	Year	Description	^{137}Cs ratio of blood to muscle	Reference
Wild boar	2012	Average of 2 juvenils	0.22	Tanoi et al. (2016).
Wild boar	2012	Average of 2 adults	0.23	Tanoi et al. (2016).
Wild boar	2013	Juvenil	0.14	Tanoi et al. (2016).
Wild boar	2013	Adult	0.41	Tanoi et al. (2016).
Pig	1960s		0.10	Green et al. (1961)
Calf	1960s		0.01	Green et al. (1961)
Cattle	2011	Average of 79 cattles (63 of adults and 13 of calves)	0.04	Fukuda et al. (2013)

(Figs. 9.5 and 9.6), indicating that the ^{137}Cs situation in the forest did not change between 2012 and 2013. However, the ^{137}Cs concentrations in organs and tissues were different between 2012 and 2013 (Figs. 9.5 and 9.6). It is unclear why stomach contents did not correlate with the organs in terms of ^{137}Cs. Furthermore, the reason for different ^{137}Cs concentrations between juveniles and adults in 2013 is unclear. We need to continue monitoring the wild boars to understand the reasons for these results.

Collecting blood is much easier than collecting other organs or tissues. Thus, it is of interest to compare the ^{137}Cs concentrations in blood with those in other organs or tissues. The ^{137}Cs concentration ratios of blood to muscles are summarized in Table 9.1. Unfortunately, the ^{137}Cs ratio of blood to muscle was not consistent, indicating that blood is not a suitable material for monitoring radiocesium concentrations in the muscle of wild boars. We also found species differences, whereby the ratios of ^{137}Cs concentration of blood to muscles in wild boar, 0.24 on average, were much higher than the value of 0.04 reported for cattle (Fukuda et al. 2013), 0.01 for calf (Green et al. 1961), and 0.10 for pig (Green et al. 1961). The reason why these ^{137}Cs concentration ratios differed so markedly among animals remains unclear.

Acknowledgements The authors thank Hidenori Ishii, Tsugio Kanno, Tsuguo Takanashi, Tadashi Yamaguchi, Mitsuro Kanno, and Gohei Hayashi for capturing wild boars; Sadanobu Abe and Koichi Sato for euthanizing wild boars; and Kazuyuki Uchida, Chiyo Doi, Naoto Nihei, Atsushi Hirose, Natsuko I. Kobayashi, Ryohei Sugita, Tatsuya Nobori, Tomoko M. Nakanishi, Muneo Kanno, Ippei Wakabayashi, Miicha Ogawa, and Yoichi Tao for collecting and measuring samples.

References

Fukuda T, Kino Y, Abe Y et al (2013) Distribution of artificial radionuclides in abandoned cattle in the evacuation zone of the Fukushima Daiichi nuclear power plant. PLoS ONE 8:e54312. doi:10.1371/journal.pone.0054312

Green RM, McNeill KG, Robinson GA (1961) The distribution of potassium and caesium-137 in the calf and the pig. Can J Biochem Physiol 39:1021–1026. doi:10.1139/o61-102

Hamada N, Ogino H, Fujimichi Y (2012) Safety regulations of food and water implemented in the first year following the Fukushima nuclear accident. J Radiat Res 53:641–671. doi:10.1093/jrr/rrs032

Ohmori H, Sasaki Y, Tajima K, Katsumata M (2014) Radioactive caesium concentrations in pigs fed brown rice contaminated by the Tokyo Electric Power Company Fukushima Daiichi nuclear power plant. Livest Sci 159:156–160. doi:10.1016/j.livsci.2013.10.026

Saito K, Tanihata I, Fujiwara M et al (2015) Detailed deposition density maps constructed by large-scale soil sampling for gamma-ray emitting radioactive nuclides from the Fukushima Dai-ichi Nuclear Power Plant accident. J Environ Radioact 139:308–319. doi:10.1016/j.jenvrad.2014.02.014

Schneider S, Walther C, Bister S et al (2013) Plutonium release from Fukushima Daiichi fosters the need for more detailed investigations. Sci Rep 3:2988. doi:10.1038/srep02988

Steinhauser G, Schauer V, Shozugawa K (2013) Concentration of strontium-90 at selected hot spots in Japan. PLoS ONE 8:e57760. doi:10.1371/journal.pone.0057760

Tanoi K, Uchida K, Doi C et al (2016) Investigation of radiocesium distribution in organs of wild boar grown in Iitate, Fukushima after the Fukushima Daiichi nuclear power plant accident. J Radioanal Chem 307:741–746. doi:10.1007/s10967-015-4233-z

Yasunari TJ, Stohl A, Hayano RS et al (2011) Cesium-137 deposition and contamination of Japanese soils due to the Fukushima nuclear accident. Proc Natl Acad Sci USA 108:19530–19534. doi:10.1073/pnas.1112058108

Zheng J, Tagami K, Watanabe Y et al (2012) Isotopic evidence of plutonium release into the environment from the Fukushima DNPP accident. Sci Rep 2:304. doi:10.1038/srep00304

Zheng J, Tagami K, Bu W et al (2014) $^{135}Cs/^{137}Cs$ isotopic ratio as a new tracer of radiocesium released from the Fukushima nuclear accident. Environ Sci Technol 48:5433–5438. doi:10.1021/es500403h

Chapter 10
Contamination of Wild Animals: Microhabitat Heterogeneity and Ecological Factors of Radioactive Cesium Exposure in Fukushima

Ken Ishida

Abstract Wildlife, mainly 69 bird (Aves) species, has been observed in Abukuma Mountains, northeastern Fukushima Prefecture, which is the most radioactively contaminated area, over the seasons since July 2011. However, it is still unclear whether the changes in the bird community have been caused by the radioactive contamination and/or by changes in human activity, adding to the natural dynamics. The aerial dose rate at the survey area was initially estimated to be more than 100 μSv/h (mainly ^{137}Cs, ^{134}Cs, and ^{131}I), which decreased in the summer of 2014 to 0.1–20 μSv/h (mainly ^{137}Cs and ^{134}Cs). Radioactivity in wild habitats is heterogeneous among the microhabitats and is dynamic through time and seasonal conditions. Microhabitat radio-heterogeneity was clearly indicated by the 2-month-long measurement with 200 dosimeter badges in this study. The ecological factors related to free living (in-situ) wildlife in the highly contaminated area of Fukushima are discussed.

Keywords Free-living • Wildlife • Microhabitat • Ecosystem • Radiation • Heterogeneity • Dynamics • Long term • Fukushima • Bush-warbler

10.1 Introduction

Monitoring the ways in which free-living animals, say in-situ wildlife, experience changing radiation levels and other environmental factors at Fukushima and Chernobyl (where the severe, IAEA rank seven, nuclear power accidents have been experienced), is crucial for the future world of higher radiation levels (Akimoto 2014; Hiyama et al. 2012; Ishida 2013; Møller et al. 2013). Since March 11, 2011, highly contaminated areas stretch northwestward from the Fukushima Daiichi nuclear power plant (F1NPP), to the northern Abukuma highland, Fukushima

K. Ishida (✉)
Graduate School of Agricultural and Life Sciences, The University of Tokyo, 1-1-1, Yayoi, Bunkyo-ku, Tokyo 113-8657, Japan
e-mail: ishiken@es.a.u-tokyo.ac.jp

T.M. Nakanishi, K. Tanoi (eds.), *Agricultural Implications of the Fukushima Nuclear Accident*, DOI 10.1007/978-4-431-55828-6_10

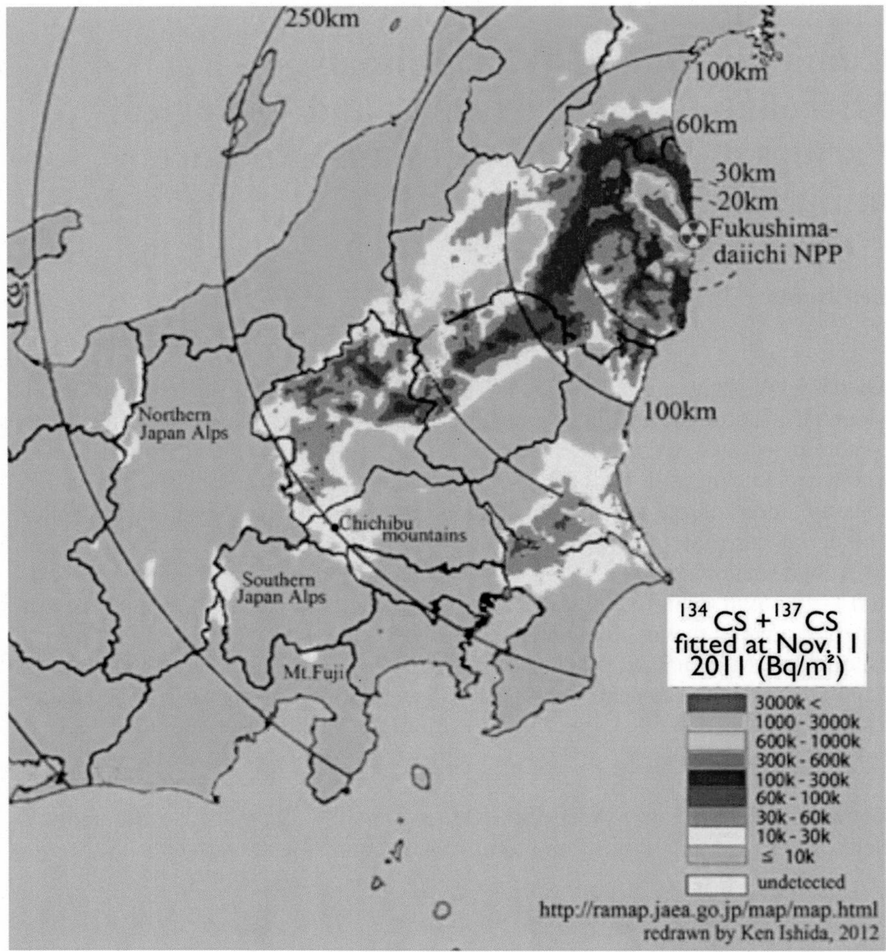

Fig. 10.1 Map of radiation fallout distribution and the location of the study area

Prefecture, Japan (Figs. 10.1, 10.2, and 10.3). The landscape of this area comprises paddy field, farmlands, natural and artificial forests, grasslands, ponds, and streams, where the biodiversity is directly related to the habitat heterogeneity. The aerial dose rates were initially estimated to be more than 100 μSv/h at some hot spots even farther than 30 km from the F1NPP. For example, the maximum value was 170 μSv/h at Techiro, Akaugi, Namie, Fukushima during spring prior to new plant growth and prior to breeding for most birds and other animals (estimated on March 16, 2011, Ministry of Education, Culture, Sports, Science and Technology, Japan (MEXT) 2011). Within the habitat, there were about 15 terrestrial mammals including the macaque monkey (*Macaca fuscata*) and about 70 birds including the bush warbler (*Cettia diphone*). I first describe the dominant bird species, habitat

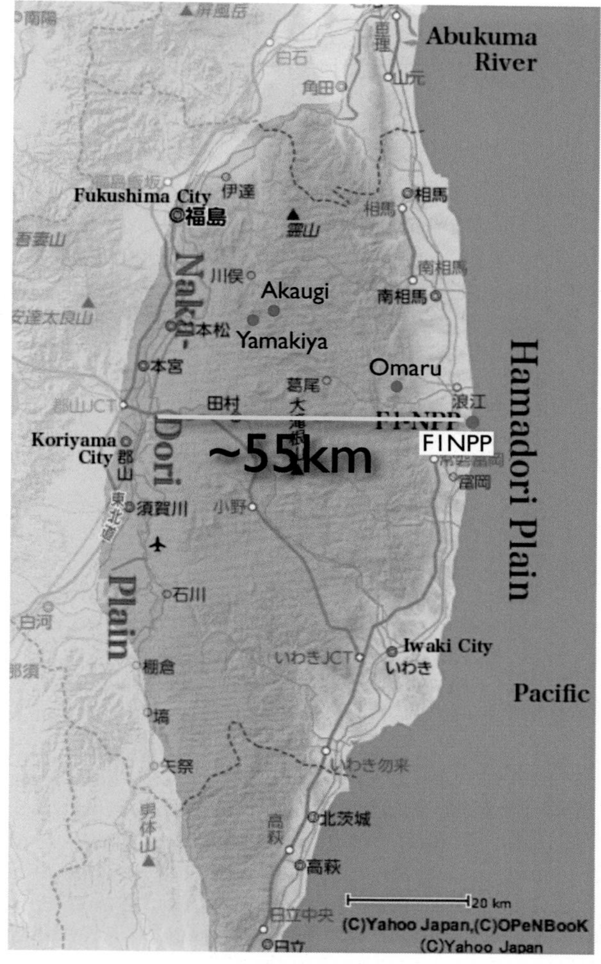

Fig. 10.2 Map of Abukuma Mountains, Fukushima Daiichi nuclear power plant (F1NPP), the three main study sites (Akaugi and Omaru of Namie Town, and Yamakiya of Kawamata Town). The *yellow line* at the center indicates the 55-km long section from F1NPP to Abukuma River, the vertical profile of which is shown in Fig. 10.3

heterogeneity through the study area, and microhabitat radiation heterogeneity in it. Then I will discuss on the analyses and future gain of long term monitoring of free-living birds.

10.2 Landscape, Climate, and Biodiversity of Abukuma Highlands

A description of the Abukuma Mountains was given in the previous volume (Ishida 2013). Here I explain the detailed characteristics of the habitats, which relate to wildlife radiation exposure and biodiversity of the most highly contaminated areas.

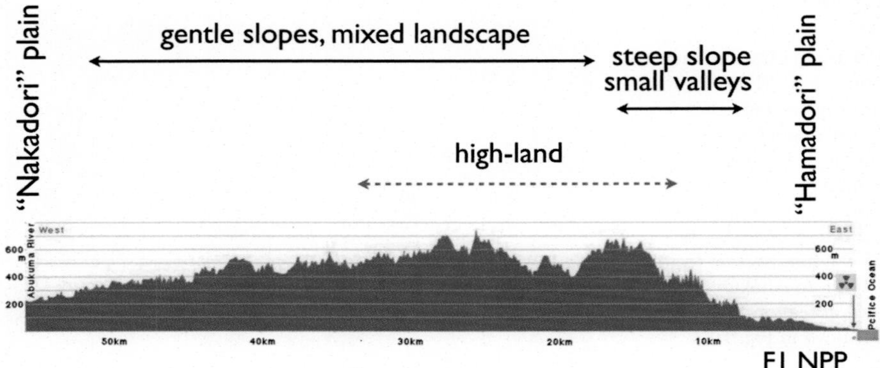

Fig. 10.3 Topological profile and landscape characteristics along the 55-km East–West line from F1NPP to Abukuma River, or Koriyama City (Refer to the *yellow line* in Fig. 10.2)

I define a "high" contamination area as that in which the estimated fallout of total ^{137}Cs and ^{134}Cs is more than 3000 kBq/m^2 (red area in Figs. 10.1 and 10.4), and "low" or "none" contamination areas in which the contamination is less than 1000 kBq/m^2 (yellow, mauve to green in Figs. 10.1 and 10.4). I explain why I do not use the measurements of dose rates recorded at each place for comparison in Sect. 10.4.

The F1NPP reactors are located on the eastern Pacific coast (Fig. 10.2). The closest to F1NPP reactors and most highly contaminated area is in the narrow flat plain of 5–10 km (Hamadori Plain). This region comprises residential areas among paddies and farmland (Fig. 10.3). Next to this plain is a steep slope with small valleys covered by a forest of coniferous plantations, secondary pine, and deciduous trees. The elevation increases quickly up to about 400 m and the slope of this part of the eastern side functions as a radiative shield. Such terrain effects were also indicated at in farther areas such as Chichibu Mountains (the southwestern edge of the contamination area expanding to 250 km from FINPP) and Iide Mountains (150 km west) as outlined in the MEXT report (2011, pp. 25–26).

The western area contains a highland plateau with many small hills about 600 m above sea level. The highest peak of Mt. Ootakine is 1192 m, the next Mt. Hiyama 1057 m, and the others are less than 1000 m. This highland plateau comprises a scattering of paddies, farmland, pastures, orchards, coniferous and deciduous forests, grasslands, streams, and ponds. Habitat heterogeneity and biodiversity are highest around this landscape. On the western side of Abukuma Mountains is a gentle slope with low hills, farmland, and forest (Fig. 10.3), which connects to the eastern highland plateau (it should be noted that there is no distinct border between these regions). Less of this area has been developed into agricultural fields as compared to the East, and more people live toward the West of Abukuma River. Along this river, there were several central cities of the Fukushima Prefecture, such as Koriyama, Nihonmatsu, and Fukushima (Nakadori Plain).

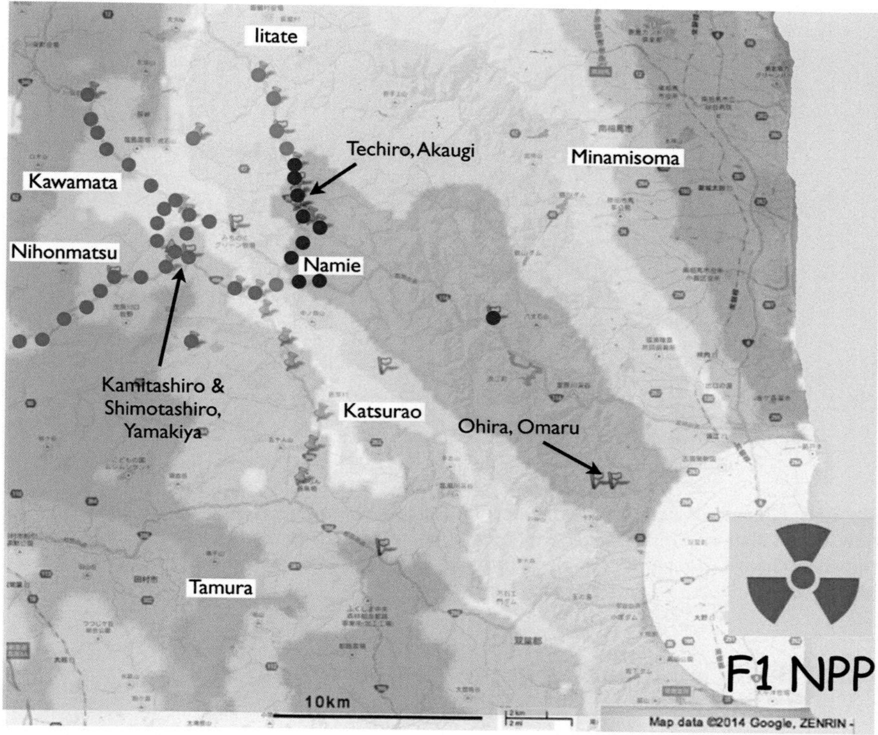

Fig. 10.4 Sites of early morning bird point count survey and microhabitat aerial dose rate measurements: *Blue dots* indicate the sites in the highest (>3000 kBq/m^2), *greens* in the middle (1000–3000 kBq/m^2), *reds* in the lower (<1000 kBq/m^2) contamination areas. Not all the sites are presented here

There are significant seasonal changes evident by the following phenomenon: rapid plant growth in spring (April to May); a flash of insects and other small animals such as frogs and spiders in summer (June to August); leaf coloring and falling in autumn (September to November); and snow accumulation in winter mainly on the highland plateau (December to March, Fig. 10.5). On March 11, 2011, there was some snow cover left on the ground in the highlands, and most plants and animals were not yet active or had not yet arrived for breeding. Large areas, except the eastern steep slope, is occupied by paddy fields, which also undergo significant seasonal changes; water flooding during the growth of rice and rapid growth of rice stems, which is then followed by the disappearance of both during autumn. The paddy fields are surrounded by streams and forests (see also Fig. 12.3 in Ishida 2013).

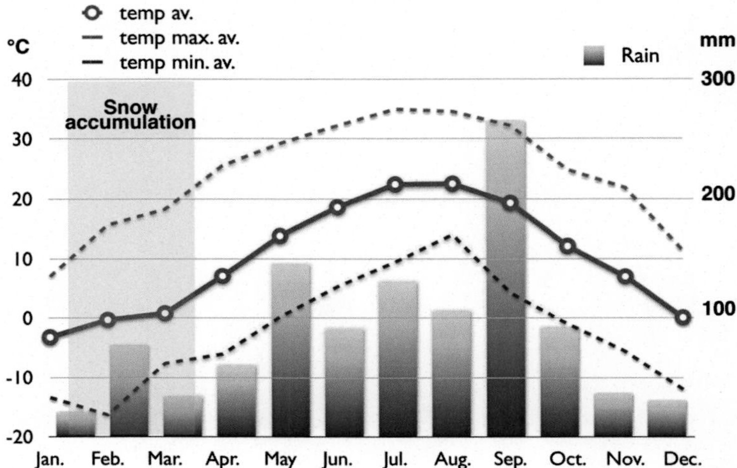

Fig. 10.5 Climate of Iitate Village, data from Japan Meteorological Agency for 2011: there is snow accumulation during January to March, more rainfall in early summer, and seasonal temperate change. Sometime typhoons (tropical cyclones) bring much rain in late summer or early autumn

10.3 Bird Communities in the Northern Abukuma Highlands

The author observed 69 bird species during 25 surveys in the study area from July 2011 to August 2014. During 5-min point count surveys (Møller et al. 2013), which were conducted in the early mornings of mid-June (3:40–9:00 a.m.), 3–13 species were recorded at each count (Figs. 10.4 and 10.6). I recorded fewer species in the high contamination area and more species in the intermediate and low radiation areas in 2012, although the differences were not statistically significant ($P > 0.05$, Mann–Whitney *U*-test). After 1 April 2013, entrance was restricted by law, and I could only perform a few early morning point counts within the area. So the range, average, and standard deviation of the number of observed species in only the low contamination area are shown for 2013 and 2014 in Fig. 10.6. These data illustrate the agreement on that the degree of diversity was lower for bird and butterfly species and higher for spiders in the high contamination areas in Fukushima (Fig. 10.3 in their paper, Møller et al. 2013). This mixed result may have arisen from of habitat diversity change, radiation, and/or human activities.

The 10 most dominant species in the point count survey were almost common during the 3 years of observed breeding seasons (Table 10.1). In addition, the three most dominant species, *Cettia diphone*, *Hypsipetes amourotis*, and *Emberiza cioides*, were also consistent in their ranking order, and four other species, *Cuculus policephalus*, *Corvus macrorhynchos*, *Parus minor*, and *Phasianus versicolor* were always in the list and have been frequently recorded. *C. policephalus* is a parasite of

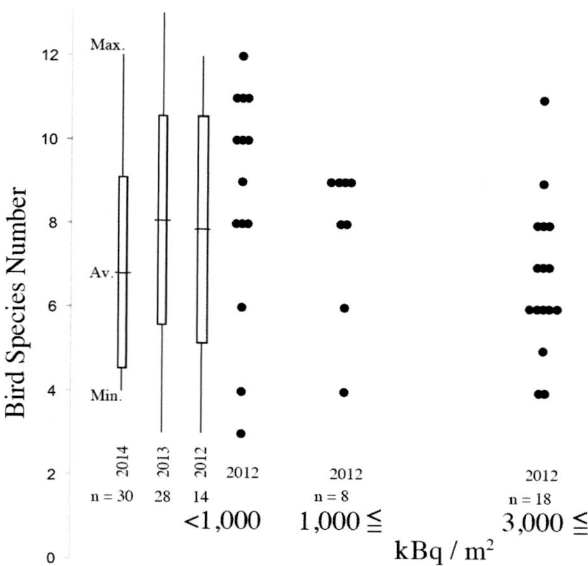

Fig. 10.6 Bird species richness estimated with early morning point count surveys in mid-June: the survey within the higher contamination area (1000 kBq/m^2) was restricted since April 2013. Average and ranges (minimum, standard deviation, and maximum) of species numbers recorded are shown for the 3 years of 2012–2013, only for the low radiation (<1000 kBq/m^2) district

Table 10.1 Ten dominant species of birds in point count survey (observation frequency rate)

Rates are per 38, 34, and 33 counts in each year					
2012		2013		2014	
Cettia dipohne	0.84	*C. dipohne*	1.00	*C. dipohne*	0.97
Hypsipetes amourotis	0.76	*H. amourotis*	0.94	*H. amourotis*	0.94
Emberiza cioides	0.55	*E. cioides*	0.68	*E. cioides*	0.58
Cuculus poliocephalus	0.47	*P. minor*	0.56	*C. macrorhynchos*	0.55
Corvus macrorhynchos	0.45	*C. macrorhynchos*	0.53	*C. poliocephalus*	0.42
Parus minor	0.47	*Streptopelia orientalis*	0.50	*Acrocephalus arundinaceus*	0.42
Phasianus versicolor	0.39	*C. poliocephalus*	0.47	*P. minor*	0.33
Carduelis sinica	0.32	*P. versicolor*	0.35	*Poecile varius*	0.27
Zosteropus japoniucs	0.26	*Passer montanus*	0.26	*P. versicolor*	0.24
Streptopelia orientalis	0.26	*Garrulax canorus*	0.24	*Passer montanus*	0.21

the most dominant *C. diphone*. Overall, these seven species are common as edge species or users of variable ecosystems. *Acrocephalus arundinaceus* is a common summer visitor that has a loud call. It inhabits marshlands and breeds in reed beds. This species may have immigrated earlier or more in 2014 than the other years, and was recorded in the point counts conducted during mid-June. These seven or eight dominant species might be good indicator birds of environmental change and/or radiation effects in this area (Gregory and Strien 2010).

Other species are less common and their populations fluctuate annually. For instance, *Passer montanus* is a common species around human houses and paddies all over Japan. *Streptopelia orientalis*, the common pigeon in low elevations in Japan, was also frequently observed much more in 2011. Pigeons and doves are strong flyers and can quickly change their distribution depending on their food resource. *Carduelis sinica* was observed at more than 20 % of point count sites in 2013, *Garrulax canorus* more in 2012, and *Zosteropus japoniucs* more in 2014. *Poecile varius* was dominant only in 2014. It will depend on the future observations as to whether these species will contribute toward change in the wildlife after the F1NPP accident.

10.4 Heterogeneity of Microhabitat Radioactivity

It has been proven that radioactive substances, ^{137}Cs and ^{134}Cs, strongly fix to clay soils in the ecosystem (Chap. 1). The purpose of many studies in Fukushima is to investigate the dynamics of Cs within and among the ecosystems. Their macro-distribution has been well examined and published on the web (MEXT 2011). It is also well known that there are hotspots and that dose rates are higher close to the ground. I examined dose rates in the field with dosimeter badges (type ES, Chiyoda Technol Co. Ltd., Tokyo) set at tree trunks for about 2 months.

The dose rates at about 1 m above the ground in forest stands decreased according to the half-life of ^{137}Cs and ^{134}Cs, and they were lower during the season with snow accumulation (Figs. 10.5 and 10.7).

I set about 100 dosimeter badges at each of the two locations to examine the microhabitat heterogeneity of radioactivity during December in 2012 and January in 2013. One location was at Techiro, Akaugi, Namie Town (37°35′33″N, 140°45′14″E) in a high contamination area, and the other was located at Kamitashiro, Yamakiya, Kawamata Town (37°34′6″N, 140°41′36 ″) in a low contamination area (Fig. 10.4). The badges were located within a 300-m radius at Techiro, and within a 200-m radius at Kamitashiro (Fig. 10.8). Most were set on the tree trunks at 0 m (on the ground), 1.2 m (breast height) and 2 m (head height). Some were set in a plastic tube at 0.15-m depth in the ground, and up to 11 or 12 m above the ground on tree trunks or on a tower (Fig. 10.8).

Higher radioactivity was recorded on the ground, and the radioactivity decreased according to the vertical distance from the ground increased (Fig. 10.9). Although only one or two samples were taken at 0.15-m depth, the radioactivity was lower than that of samples on the ground surface at the same places. For samples measured on a deciduous tree trunk, the radiation decreased in a non-linear fashion relative to the distance from the ground (Fig. 10.8, right); probably because the radioactive substances remained either in the bark, or in the moss or lichen on the bark. The dose rates heterogeneity was high among the microhabitats within a small

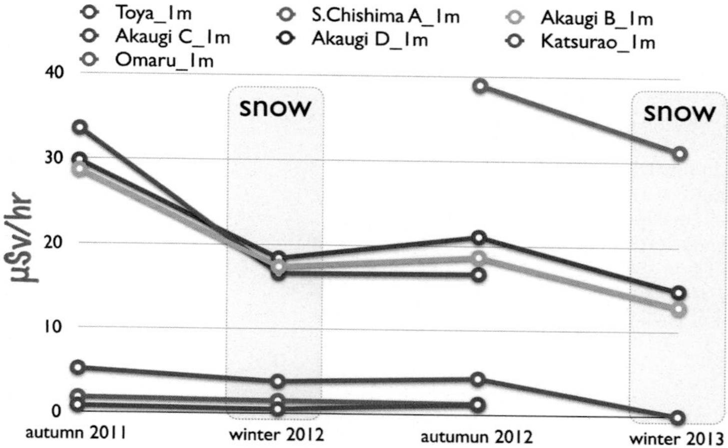

Fig. 10.7 Ecological decline process of aerial dose rate at several contaminated districts in Fukushima

Fig. 10.8 Outline profile of aerial dose rate microhabitat heterogeneity, which depends the distance (height) from the soil (ground), direction towards F1NPP, surrounding topology, vegetation, and artificial management. (**a**) Aerial view taken from the southwest with a motor hung glider: some main microhabitats and their dose rates at Kamitashiro, Yamakiya of Kawamata Town (F1NPP is right forward), (**b**) an oak tree and its dose rates along its trunk (0 to about 11 m from the ground) at Techiro, Akaugi of Namie Town. The dose rates on the ground was 12.4 (μSv/h) and at the height of 11 m was 10.5 (μSv/h). See also Fig. 10.9

area. The dose rates in Kamitashiro and in Techiro varied tenfold and threefold, respectively (Fig. 10.9). The dose rates depended on the direction of the slopes in which each microhabitat located with respect to F1NPP and on a small scale (several meters) and large scale geometric structures (Fig. 10.8, left).

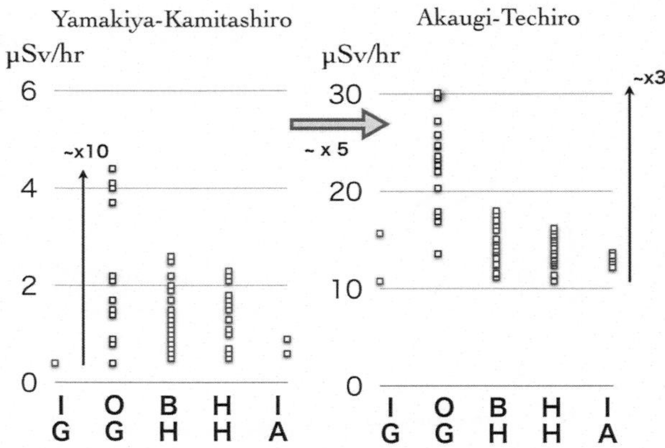

Fig. 10.9 Microhabitat heterogeneity of aerial dose rate at two sites, estimated with dosimeter glass badges (type ES, Chiyoda Technol Co. Ltd., Tokyo) fitted for 2 months in the field. Abbreviations: *IG* in the ground (15-cm depth), *OG* on the ground, *BH* breath height (1–1.2 m above the ground), *HH* head height (~2 m), *IA* in the air (>3 m)

10.5 Effects of Radioactivity on Bush Warbler and Boar

The bush warbler (*Cettia diphone*) is the most dominant bird, and its microhabitat is the dense bush at lower heights, usually up to 2 m from the ground, in a variety of vegetation and ecosystems from the coast to the mountaintop. Wild boar (*Sus scrofa*) is a common mammal in the forest and grasslands, and Abukuma highland is the northernmost border of its distribution in Japan. The boar is omnivorous and often forages for various things by digging into the soil with its strong nose. Therefore, both of these animals have a high probability of being close to the radioactive substances in the contaminated area of Fukushima, and should be two of the best indicator species to monitor the effects of the F1NPP accident on free-living animals.

Tail feathers (Pirastro et al. 1993) of the bush warblers, caught at Akaugi, were contaminated with up to 556 Bq/g of ^{137}Cs, ^{134}Cs, and ^{110m}Ag in mid-August 2011 (Ishida 2013). The contamination level of feathers at the same site decreased to about 20 % in 2012 and contamination became less frequent among the individuals in 2013. Fukushima Prefecture (2014) published radioactivity measurements of the muscle of game animals, including boar, on its website. For this study, I picked up a total of 680 samples collected from boars, caught in Fukushima during the period of 22th May 2011 until 20th August 2014. The dose rates ranged from “not detected” (^{137}Cs < 5.5 Bq/kg and ^{134}Cs < < 4.0 Bq/kg) to 61 Bq/g (a boar from Minamisoma City on 11 March, 2013). The highest value in 2014 was 2.4 Bq/g (a boar from Minamisoma City on 27 July). Wildlife exposure measurements started soon after the F1NPP accident, and long term monitoring of these animals may contribute to our understanding the dynamics of radioactive substances in the ecosystem (and evnvironmental protection; ICRP 2008).

10.6 Environmental Factors that Affect Wildlife

Microhabitat heterogeneity of radioactivity was as large as tenfold (Fig. 10.9). The long, low, and internal exposure to radiation (ECRR 2010) by wildlife depends on the species, on the behavior of individual animals, and on where and when it is present at which microhabitat. The difference in microhabitat radiation caused by the direction with respect to F1NPP can affect these animals incidentally. It is first necessary to determine the individual exposure as much as possible or to sample as many individuals as possible to account for the individual variation. Environmental factors, which were caused by F1NPP accident and thought to relate to wildlife fitness change, include radioactivity as well as the numerous interactive changes. Both of these environmental factors have a direct and indirect effect, and sometimes their effects cascade through food webs (Hallmann et al. 2014). At least in the early years of 2011 and 2012, there seemed to be a notable decrease in butterfly fitness in the contaminated areas of northern Abukuma highland (Hiyama et al. 2012; Møller et al. 2013). I tried to catch butterflies at highly contaminated areas such as at Akaugi and Omaru (Fig. 10.4) during the 2012 summer, but I could not find any. Then, I found some in the summer of 2013, and more in 2014. Butterflies and some other insects are an important food resource for insectivorous birds, too. Human activity, including cultivation of the agricultural land, cutting timbers, hunting, and collecting woods, mushrooms and bamboo shoots, has stopped in the high contamination areas and in some parts of the low contamination areas. Most of the low contamination areas, where people are expected to return to live in several years, were decontaminated in an extreme manner. The soil and the lower vegetation were removed. This resulted in a loss of several microhabitats and biodiversity at these sites. In high contamination areas, where people are not expected to return for at least several decades, the land has been left without human activity and some wildlife has increased in density, because of the lack of hunting by humans. All of the complex interactions in an ecosystem should affect the bird community, and we should consider and detect more than the few most important factors.

Cettia diphone, for example, may have some advantage by inhabiting dense bush at the abandoned agricultural land and secondary forest edge for a while until the forests regrown, and some disadvantage by the fewer hymenopteran insects (Hallmann et al. 2014) and the removal of the bush as part of the extreme decontamination activity. As *C. diphone* stays close to the ground, it could be affected by the radiation at least over the first couple of years, while the dose rate in the air is rather high. *Hypsipetes amourotis* is an arboreal bird and is active within the forest canopy. It is omnivorous and has a large home range; hence, the various environmental changes could be neutral for this bird. *Emberiza cioides* is a seed-eater and is a stable settler in its small territory. It often feeds on seeds of shorter grasses and also on the ground, so its exposure situation resembles that of *C. diphone*, except for their food feeding habits. I currently do not have enough information about seed production in the highly contaminated area, but as the F1NPP accident occurred prior to the rapid growth of plant shoots in spring, seed production seemed to be less affected by the radiation.

Phasianus versicolor is a ground dwelling bird and is omnivorous. It forages and nests on the ground. It is a game bird. There should be multiple responses by this species to the environmental changes. Unfortunately, there is very little information on the wildlife in the contaminated area of Abukuma Mountains before the F1NPP accident. It will be desirable to monitor the ecosystem and several principal wildlife species, such as some of the dominant 10 birds (explained previously), *S. scrofa* and unique primate *Macaca fuscata* (Hayama et al. 2013), until most of the ecosystems and human societies regain stability in five or more decades later (Galvan et al. 2014).

References

Akimoto S (2014) Morphological abnormalities in gall-forming aphids in a radiation-contaminated area near Fukushima Daiichi: selective impact of fallout? Ecol Evol 4:355–369

European Committee on Radiation Risk (2010) In: Busby C (ed) Recommendations of the European committee on radiation risk. The health effects of exposure to low doses of ionizing radiation, p 248. http://www.euradcom.org/2011/ecrr2010.pdf

Fukushima Prefecture (2014) Data of wildlife radiation monitoring (in Japanese). http://www.pref.fukushima.lg.jp/sec/16035b/wildlife-radiationmonitoring1.html

Galvan I, Bonisoli-Alquati A, Jenkinson S, Ghanem G, Wakamatsu K, Mousseau SA, Møllor AP (2014) Chronic exposure to low-dose radiation at Chernobyl favours adaptation to oxidative stress in birds. Funct Ecol 28:1387–1403

Gregory RD, van Strien A (2010) Wild bird indicators: using composite population trends of birds as measures of environmental health. Ornithol Sci 9:3–22

Hallmann CA, Foppen RPB, van Turnhout CAM, Kroon H, Jongejans E (2014) Declines in insectivorous birds are associated with high neonicotinoid concentrations. Nature. doi:10.1038/nature13531

Hayama S, Nakiri S, Nakanishi S, Ishii N, Uno T, Kato T, Konno F, Kawamoto Y, Tsuchida S, Ochiai K, Omi T (2013) Concentration of radiocesium in the wild Japanese monkey (*Macaca fuscata*) over the first 15 months after the Fukushima Daiichi disaster. PLoS ONE. doi:10.1371/journal.pone.0068530

Hiyama A, Nohara C, Kinjo S, Taira W, Gima S, Tnahara A, Otaki JM (2012) The biological impacts of the Fukushima nuclear accident on the pale grass blue butterfly. Sci Rep 2:570. doi:10.1038/srep00570

ICRP (2008) Environmental protection – the concept and use of reference animals and plants. http://www.icrp.org/publication.asp?id=ICRP%20Publication%20108

Ishida K (2013) Contamination of wild animals: effects on wildlife in high radioactivity areas of the agricultural and forest landscape. In: Nakanishi TM, Tanoi K (eds) Agricultural implications of the Fukushima nuclear accident. Springer, Tokyo, pp 119–129. doi:10.1007/978-4-431-54328-2_12

MEXT (2011) Press release on the airborne survey results 1910-1125_2, p 27 (in Japanese). http://radioactivity.nsr.go.jp/ja/contents/5000/4900/24/1910_1125_2.pdf

Møller AP, Nishiumi I, Suzuki H, Ueda K, Mousseaue TA (2013) Differences in effects of radiation on abundance of animals in Fukushima and Chernobyl. Ecol Indic 24:75–81

Pirastro A, Congiu L, Tallandini L, Turchetto M (1993) The use of bird feathers for the monitoring of cadmium pollution. Arch Environ Contam Toxicol 24:355–358

Chapter 11
Translocation of Radiocesium in Fruit Trees

Daisuke Takata

Abstract We report our findings on the translocation and distribution of radiocesium inside fruit trees after radiocesium (Cs) was released by the accident at the Fukushima Daiichi Nuclear Power Plant.

We examined the differences in the rooting depth of grapes and figs and the translocation of radiocesium from the soil to the plants. Much of the radiocesium fallout from the nuclear accident remained on the surface soil layer; however, in environments such as orchards, radiocesium translocates more easily to aboveground parts of trees with shallower roots than of those with deeper roots. It was observed that if the old branches were the source of radiocesium, translocation occurred to both the new aboveground organs, such as leaves and fruits, and the underground parts, including pioneer roots. It was clarified that translocation from old organs contributed a much higher proportion of accumulated radiocesium to fruits than that from the soil. We reported that immediately after the accident, radiocesium that accumulated on the bark quickly infiltrated inside the trees. However, several months after the accident, it is possible that a decreased proportion translocated from the outer bark to the internal parts of trees, such as the wood. The translocation of radiocesium and potassium (K) into fruits and leaves may show some differences. Explaining the behavior of radiocesium translocation in perennial crops using K as an index is even more difficult than that in annual crops. To predict the radiocesium concentrations in the harvested fruits, the concentrations in the thinned fruits and the harvested fruits were compared. The results showed that there is a strong correlation between the two. However, since some trees were outliers, predictions must be made carefully.

Keywords Fig • Fruit tree • Grapes • Peach • Radiocesium

D. Takata (✉)
Institute for Sustainable Agro-ecosystem Services, Graduate School of Agricultural and Life Science, The University of Tokyo, 1-1-1 Midori-cho, Nishi-tokyo-shi, Tokyo 188-0002, Japan
e-mail: takata@isas.a.u-tokyo.ac.jp

T.M. Nakanishi, K. Tanoi (eds.), *Agricultural Implications of the Fukushima Nuclear Accident*, DOI 10.1007/978-4-431-55828-6_11

11.1 Introduction

Major fruit production in Fukushima Prefecture in 2013 was cultivated over 5762 ha and produced a crop yield of 84,060 t; in terms of cultivation area, Fukushima ranked 9th in Japan (Ministry of Agriculture, Forestry and Fisheries, Japan (MAFF)). When itemized, the crop yield of peaches ranked 2nd (share 23 %, 124,700 t, 2013). This shows that, even after the nuclear accident, Fukushima was still a major producer of fruits in Japan (Table 11.1). Many crops did not suffer significant reductions to their growing area and yield; however, the shipment volume of persimmons decreased to 1/3 in the year of the accident, and has not changed much since then (Table 11.2). However, even crops that did not suffer decreased shipping volumes were significantly impacted. For example, there was an increased shipment of crops, such as peaches through agricultural cooperatives, to the market during the year of the accident. The following factors are believed to have contributed. Sales were poor due to the consumers' concerns over radiocesium, indicating that direct sales and orchard sales to tourists decreased. As a response to consumer reactions, producers changed the sales policy. Together, these factors led to an increase in the stock volume; thus, in the year of the accident, the wholesale price

Table 11.1 Shipments, crop yields and cultivated area of Fukushima Prefecture in 2013 (from Ministry of Agriculture, Forestry and Fisheries, Japan 2013)

Fruit trees	Ranked	Cultivated area (ha)	Crop yields (t)	Shipments (t)
Peach	2nd	1780	29,300	27,100
Apple	6th	1380	26,800	23,500
Persimmon	5th	1340	3790	3790
Japanese pear	4th	974	19,800	18,300
Grapes	13th	288	2960	2960

Table 11.2 Changes in shipments and crop yields and cultivated area of peach and persimmon in Fukushima Prefecture (from Ministry of Agriculture, Forestry and Fisheries, Japan 2013)

	Peach			Persimmon		
Year	Cultivated area (ha)	Crop yields (t)	Shipments (t)	Cultivated area (ha)	Crop yields (t)	Shipments (t)
1990	2610	40,200	37,600	1350	10,200	8270
2000	1800	31,500	29,400	1420	11,000	8670
2004	1750	30,700	28,400	1430	13,500	11,200
2008	1790	31,800	29,400	1420	16,100	13,500
2009	1790	30,100	27,900	1410	13,600	11,300
2010	1780	28,200	26,200	1400	14,000	12,000
2011[a]	1780	29,000	26,500	1390	4550	3540
2012	1780	27,500	25,200	1350	4480	3390
2013	1780	29,300	27,100	1350	4890	3790
2013 in Japan	10,700	124,700	114,100	22,300	214,000	177,400

[a]Accident year

Table 11.3 Stock volume, Market share and wholesale price of monthly Fukushima Prefecture peach in Tokyo metropolitan central wholesale market (from Monthly report of Tokyo metropolitan central wholesale market, 2013)

	August			September			August–November Total/average		
Year	Stock volume (t)	Market share (%)	Wholesale price (Yen/kg)	Stock volume (t)	Market share (%)	Wholesale price (Yen/kg)	Stock volume (t)	Market share (%)	Wholesale price (Yen/kg)
2008	2837	43.0	359	755	45.2	408	3599	43.3	369
2009	2751	48.5	384	537	43.3	411	3294	47.5	389
2010	2421	42.5	439	868	41.9	481	3309	42.2	450
2011[a]	4631	54.1	196	786	38.3	300	5433	50.9	211
2012	2834	48.4	344	998	42.0	364	3881	46.3	350
2013	3916	61.5	342	622	36.7	421	4574	56.0	353

[a]Accident year

dropped when compared to the previous year (Table 11.3). However, the stock volume returned to the pre-accident value in the following year, leading to a certain level of recovery in the wholesale price. Prices have shown some stability in the years following the accident.

However, compared to other crops, such as rice, there have not been many studies conducted on radiocesium translocation in fruit trees since the nuclear accident. Thus, there are still many factors that need to be determined. In a previous publication (Takata 2013), the impact of radiocesium on the translocation and distribution inside fruit trees was investigated. The behavior of radiocesium inside fruit trees after the nuclear accident differed to that found after the Chernobyl accident. It was thought that significant effects occurred because at the time of the accident many deciduous fruit trees had not yet flowered or begun producing leaves. Although the accident occurred prior to the growth of leaves and fruits, radiocesium was detected in fruits at the time of harvest. In the year of the accident, the contribution of radiocesium fallout in soil to translocation from the roots was much lower than from aboveground parts. Therefore, we considered the possibility that radiocesium adhered to aboveground organs, such as the trunk and primary scaffold limbs, translocated to the inside of trees via direct deposition on old organs in aboveground parts. Translocation pathways to fruits are most likely due to translocation from old aboveground organs and absorption from the soil through the roots. However, the relative contribution of these two pathways is unclear. In addition, in single annual crops, the behaviors of radiocesium and K showed a strong relationship. Thus, the behavior of K can be a measure of translocation; therefore, application of K to reduce translocation is being undertaken. However, it is difficult to examine all the radiocesium behaviors in relation to K in perennial fruit trees, particularly when radiocesium is already present inside the trees.

In this chapter, we focus on the report from the previous publication in order to discuss the relationship between radiocesium and K in terms of the behavior of radiocesium inside the trees and in orchards.

11.2 Pathways of Radiocesium in Fruit Trees

11.2.1 Absorption from Soil

As the distribution of radiocesium in soil after nuclear accident was defined, it became clear that there was a large proportion of radiocesium in the surface layer of soil in orchards (Takata et al. 2012c, d). We demonstrated that, even in this condition, the amount of radiocesium translocated from the soil to mature peach trees in the year of the accident was minimal compared to the amount directly absorbed from aboveground parts (Takata et al. 2012b). The survey of transfer factors of fruit trees, conducted after the Chernobyl accident, showed that this does not mean that there is no translocation from the soil in orchards (IAEA 2003). Furthermore, it is possible that a large amount of radiocesium may be absorbed from the soil surrounding small (young and recently transplanted) trees, with a rapid increase in dry matter.

Among the tests to determine the transfer factors affected by the Chernobyl accident, a survey of the planting area was conducted in undisturbed areas, while reproduction tests in pots were conducted in soil with uniform radiocesium concentrations. The differences in the conditions may be the cause for the large discrepancy in transfer factors, even in the same tree species. In vegetable cultivation, radiocesium adhering to the soil surface is stirred during tilling (Oshita et al. 2013). However, orchard soil is not tilled, so it is thought that homogenization with the lower soil layer does not occur. In fruit trees, the distribution of roots with depth varies depending on the tree species, variety, and rootstock variety. Therefore, it is important to clarify the level of radiocesium translocation to fruit trees taking into consideration the variations in root distribution and heterogeneity of the soil pollutant. However, no such report is available. Therefore, we investigated translocation of radiocesium inside fruiting trees, shallow-rooted figs and deep-rooted grapes, located in an area where radiocesium was present only in the soil surface or in the lower soil layer, and we compared the organ volume distribution with naturally-occurring ^{40}K.

We tested 3-year-old ‘Campbell Early’ and ‘Muscat Bailey A’ grapes (*Vitis. spp.*), and 2-year-old ‘Houraishi’ figs (*Ficus carica* L.) in 4.0 L pots. These plants had been cultivated in a closed, environmentally controlled greenhouse before the nuclear accident. In March 2012, the trees were replanted into 7.0 L pots, and the radiocesium concentrations in the soil were adjusted. To adjust the radiocesium concentration in the soil, we used orchard soil (sandy loam) from Date, Fukushima Prefecture, and Tokyo, and uncontaminated commercial soil (Akadama soil and mulch). For example, for ‘Campbell Early,’ sections with a uniform concentration

Table 11.4 ^{137}Cs concentration in pot soil (from Takata et al. 2013b)

	Soil ^{137}Cs concentration (Bq/kg – Dry weight)			Soil ^{137}Cs content (Bq/pot)
Cultivar	0–5 cm depth	5–15 cm depth	Homogeneous	
Grapes				
Campbell early				
0–5+	833.9 ± 1.9	4.7 ± 0.0	258.8 ± 1.8	655.3
5–15+	2.0 ± 0.0	938.7 ± 1.7	618.8 ± 0.0	1543.6
Homogeneous	576.4 ± 7.1	609.1 ± 0.7	592.2 ± 5.8	1665.1
Muscat bailey A				
0–5++	1006.5 ± 0.8	2.0 ± 0.0	293.4 ± 18.7	765.9
0–5+	467.4 ± 1.1	2.0 ± 0.0	154.2 ± 9.9	390.3
5–15+	2.1 ± 0.2	483.5 ± 0.1	307.5 ± 10.0	746.7
Fig				
Houraishi				
0–5+	4068.6 ± 131.1	3.9 ± 0.6	1450.3 ± 51.5	3521.4
5–15+	1.5 ± 0.6	4208.1 ± 138.2	2818.4 ± 103.0	6355.4

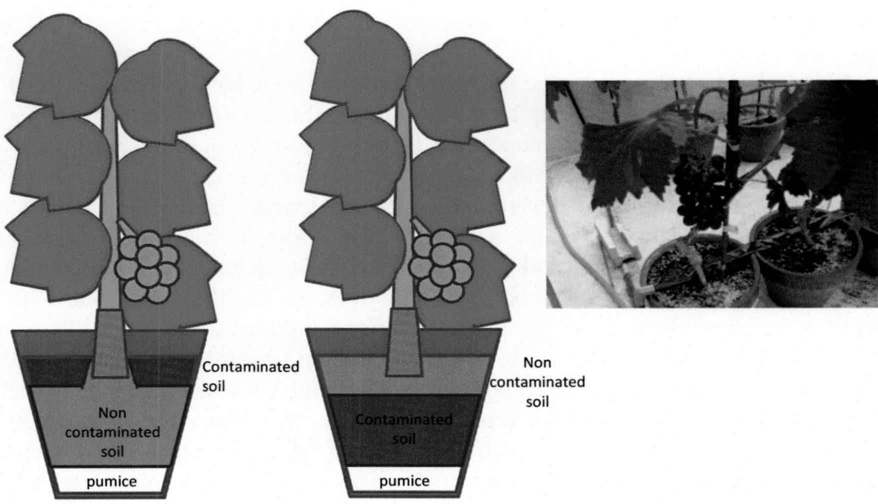

Fig. 11.1 Test of heterogeneity of ^{137}Cs concentration in pot soil. *Left*: 0–5+, *right*: 5–15+, Photograph: the image of potted grape tree

were established: a high concentration of radiocesium in the soil 0–5 cm (surface layer; 0–5+), and the 5–15 cm (lower layer; 5–15+) (Table 11.4, Fig. 11.1). At 5 cm, where the concentration changed, three layers of water-soluble packaging paper were placed to delineate the boundary. Similar treatment sections were established for the 'Muscat Bailey A' grape and 'Houraishi' fig. The concentration of ^{40}K in the soil was about 300 Bq/kg DW. After cultivation, the trees were divided into fruits, leaves, shoots, old branches (1–3-year-old branches, including rootstock), and roots. The dry weight was measured and concentrations of ^{137}Cs and ^{40}K were measured. In the two varieties of grapes and figs, the dry weight of roots from 0 to 5 cm was lower than that from 5 to 15 cm (Table 11.5). The proportion of 0–5 cm roots was higher in figs than in grapes (Fig. 11.2). The weight of the 0–5 cm pioneer roots of grapes was less than that from 5 to 15 cm. However, for figs, it was higher than that for 5–15 cm roots. Based on these results, we conclude that fig roots grew more vigorously at shallower depths than grape roots, so these plants would be suitable for comparing the effect of root distribution with depth on radiocesium uptake.

After converting the concentration of ^{137}Cs in fruits to per fresh weight (Bq/kg FW), the value was divided by the concentration in the soil (Bq/kg DW) to obtain the transfer factor. The ^{137}Cs concentration in the soil was obtained by homogenizing the soil from 0 to 5 and 5 to 15 cm depths unless the soil was uniformly contaminated. In addition, the pioneer root weight was obtained by subtracting the root weight of individual roots, which were dissected separately prior to the test, from the root weight after cultivation (Table 11.6). Concentrations of ^{137}Cs in the fruits of grapes and transfer factors were higher in 5–15+ than in 0–5+, and it was

Table 11.5 Root weight in each depth of grape vine and fig trees (from Takata et al. 2013b)

	Total root weight (gDW)		New root weight (gDW)[a]	
Cultivar	0–5 cm depth	5–15 cm depth	0–5 cm depth	5–15 cm depth
Grapes				
Campbell early				
0–5+	7.65	25.05	3.85	5.18
5–15+	7.92	26.27	3.87	5.23
Homogeneous	7.68	26.88	2.73	4.85
Muscat bailey A				
0–5++	7.48	26.75	2.53	3.35
0–5+	7.63	27.23	2.60	2.73
5–15+	7.95	26.80	2.73	3.58
Fig				
Houraishi				
0–5+	15.43	24.75	7.20	5.55
5–15+	14.93	24.40	7.18	4.35

[a]New root weight = root weight after harvest – root weight before budbreak

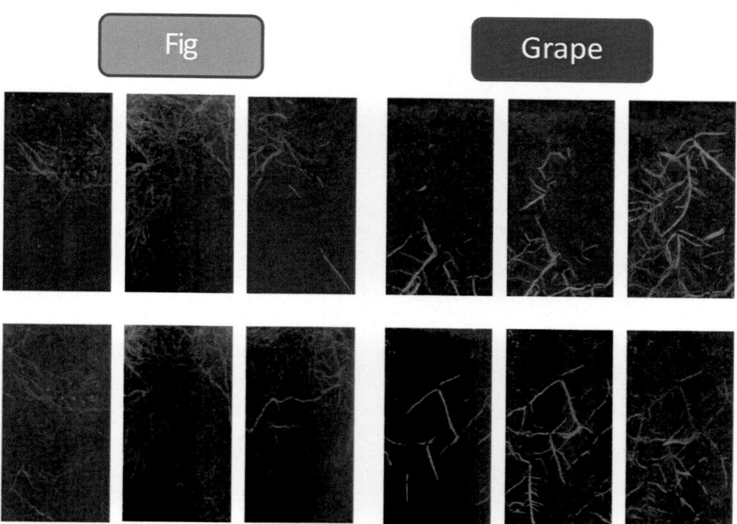

Fig. 11.2 Rooting of acrylic potted grape vine and fig trees. Roots of fig spread only upper part, and roots of grape spread all around

thought that ^{137}Cs translocation to new organs, such as fruits, derived more ^{137}Cs from the lower soil layer. In the 5–15+ section of 'Campbell Early' the concentration of ^{137}Cs in the roots was high in the lower soil layer with a high ^{137}Cs concentration. However, in other treatment sections, the concentrations in the roots were not necessarily high in soil layers with high concentrations. This indicates that the ^{137}Cs absorbed by the roots was also accumulated in the pioneer

Table 11.6 ^{137}Cs concentration in each organ and TF of grape vine and fig trees (from Takata et al. 2013b)

	^{137}Cs (Bq/kg – Dry weight)						
Cultivar	Fruit	Leaf	Shoot	Old branch	Root at 0–5	Root at 5–15	TF[a]
Grapes							
Campbell early							
0–5+	1.9 ± 0.1	3.5 ± 0.5	2.0 ± 0.5	5.2 ± 1.4	6.4 ± 0.9	6.1 ± 1.0	1.68E-03
5–15+	10.6 ± 0.4	9.7 ± 0.6	4.2 ± 0.6	5.9 ± 1.7	6.6 ± 0.9	11.1 ± 1.1	3.97E-03
Homogeneous	6.8 ± 0.2	3.7 ± 0.5	3.8 ± 0.6	5.7 ± 1.3	9.1 ± 1.1	9.1 ± 1.0	2.67E-03
Muscat bailey A							
0–5++	3.7 ± 1.1	8.4 ± 0.3	3.2 ± 0.2	7.2 ± 1.9	19.0 ± 1.8	17.5 ± 2.0	2.93E-03
0–5+	2.9 ± 0.9	6.5 ± 1.6	2.4 ± 0.2	5.8 ± 1.7	14.4 ± 2.1	13.5 ± 1.3	4.39E-03
5–15+	7.5 ± 1.0	8.1 ± 0.3	3.9 ± 0.3	7.6 ± 2.1	13.3 ± 1.6	16.1 ± 1.7	5.67E-03
Fig							
Houraishi							
0–5+	234.2 ± 23.3	195.6 ± 19.2	100.2 ± 1.2	48.0 ± 3.5	78.9 ± 5.8	41.5 ± 3.4	2.66E-02
5–15+	121.1 ± 9.5	124.4 ± 9.0	50.3 ± 10.0	21.0 ± 2.2	39.3 ± 3.9	62.0 ± 6.4	7.08E-03

[a]TF = (fruit Bq/kgFW)/(soil Bq/kgDW)

roots at different depths. This can occur in orchards, indicating that radiocesium present in the surface soil layer was absorbed by roots in the surface layer, and was translocated to roots in the lower layer. The proportion is unclear, so the rate at which this occurs is unknown. However, it is likely that this increases the concentration of radiocesium in the lower soil layer.

When contamination was either limited to the lower soil layer or was spread uniform, the problem was not substantial, because in the current situation, the contamination was limited to the surface soil layer. On the other hand, this could be a problem during planting or transplanting. For example, at the time of transplanting, by developing the planting area using soils from other locations without removing the contaminated surface, there is a risk of contaminating the lower soil layer. Fruit trees with deep roots can develop high transfer factors when the lower soil layers are contaminated; therefore, absorption from the soil may pose a critical problem compared to the situation when only the surface layer is contaminated. On the contrary, there are reports indicating that the transfer factors were high when only the surface soil layer had high concentrations compared to when the soil is homogenized. A previous report (IAEA 2003) using pots demonstrated that transfer factors in the same tree species were higher under uniform concentrations compared with the actual planting area. The issue must be sufficiently examined.

Transfer factors in the 0–5+ section for figs were similar to the values reported after the Chernobyl accident (Marouf et al. 1992). However, the values were larger than those for the two varieties of grapes. We believe that our test can reproduce the results of obtained by examining the actual planting areas, because it reproduces the significant contamination levels in the surface soil layer. On the other hand, the concentration of ^{137}Cs in new organs in the 5–15+ section was lower than that in the 0–5+ section (Table 11.4). Since such variation was noted in the 0–5+ section of figs, although the total ^{137}Cs volume in the soil was lower than that in the 5–15+ section, it is thought that in figs the ratio of ^{137}Cs translocated to new organs, such as fruits, was derived from the ^{137}Cs present in the surface soil layer which was high. This is also evident from the transfer factors. Transfer factors for the 5–15+ section had values one order of magnitude lower than the 0–5+ section. It is believed that the following aspects are related to the differences in the transfer factors of figs caused by the differences in the depth of the contaminated soil: (1) the distribution of fig roots was shallow, (2) the weight of roots in the surface layer was higher than that for grapes, and (3) the weight of pioneer roots was high (Table 11.5). In other words, the amount of rooting of pioneer roots, which play a major role in element absorption in the soil, was low in the lower soil layer, and high in the surface soil layer, leading to the different absorption of ^{137}Cs.

The changes in ^{137}Cs and ^{40}K were calculated in the organs of two varieties of grapes. These were obtained by multiplying the nuclide concentration by the organ volume to find the content per organ, then subtracting the pre-test (pre-sprouting) nuclide content from that value. In this test, each test subject was cultivated in a closed space before the accident, and had not been exposed to the fallout. In addition, ^{137}Cs was measured after homogenizing all the parts in another sample under the same conditions prior to the test, and the results were below the detection limit. Therefore, the ^{137}Cs content prior to the test was assumed to be 0. The

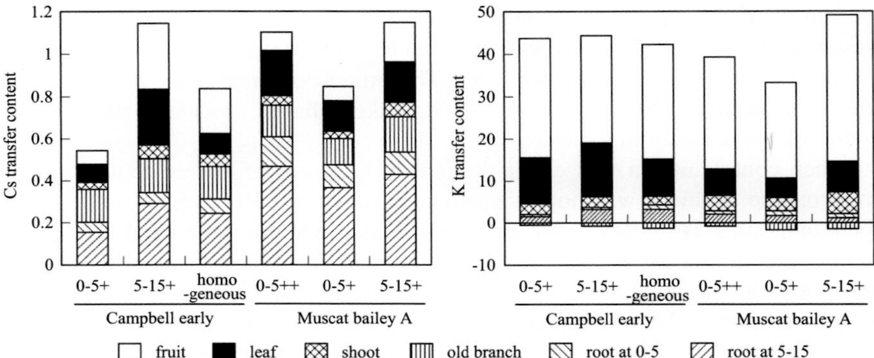

Fig. 11.3 ^{137}Cs and ^{40}K transfer content of grapevine (from Takata et al. 2013b). Transfer content = radioisotopes content *after* harvest-radioisotopes content *before* budbreak. Radioisotopes content = weight per each organ (gDW) × Bq/kgDW at each time

changes in ^{137}Cs in organs showed the same trend as in the ^{137}Cs concentration previously discussed. This indicates that there was no difference in tree growth due to the different treatments. There was a 60 % or more increase in ^{40}K in fruits for each treatment. Evidently, the changes in ^{40}K and ^{137}Cs were notably different. In this test, the change in ^{40}K in the old branches was negative (Fig. 11.3). This is thought to be related to the fact that K in the old branches was quickly used to build new organs after sprouting. On the other hand, at the time of the test, ^{137}Cs was not present inside the trees, and was not used in the early growth. Because of this, in fruits that have a large K requirement, an extreme difference between ^{137}Cs and ^{40}K occurred. The difference in the rates of ^{137}Cs and ^{40}K in fruits may be affected by the absorption from the soil and by differences in translocation to the aboveground parts. However, this is unclear based on the categories of this test.

Compared to other tree species, it is believed that the radiocesium absorbed by shallow-rooted species is high when the ^{137}Cs concentration in the surface soil layer is high. This seems to be relevant in detecting radiocesium in blueberries in Fukushima and Ibaraki prefectures. However, trees in the outdoor planting area already have ^{137}Cs internally, so the behavior of radiocesium cannot be explained simply based on translocation from the soil. Thus, in order to determine the behavior of radiocesium in perennial fruit trees, it is necessary to examine the radiocesium translocation in the old organs, and not just in the soil.

11.2.2 Translocation from Aboveground Old Organs

In the previous section, we discussed translocation of radiocesium from the soil. However, it is necessary to examine whether the contamination source of radiocesium in fruits may already be present inside the trees. Based on the test that used trees where only the soil surface was covered prior to the accident, the amount of radiocesium that was translocated from the underground to aboveground parts after the accident was so small that it was barely detectable. It is clear that the majority of radiocesium was directly absorbed as it adhered to the tree at the time of the fallout. Furthermore, when contaminated peach trees were planted and cultivated in uncontaminated soil, radiocesium translocated from old organs to new organs, including fruits. The distribution inside the trees showed a high concentration in the primary scaffold limbs and the main trunk. Investigating how the radiocesium present in the old branches of trees is translocated inside the trees at the time of cultivation is just as important (or more so) as investigating translocation from the soil. To study this aspect, tests have been conducted by painting leaves with radiocesium (Zehnder et al. 1995; Carini et al. 1999); however, there have been no reports on the translocation of radiocesium from branches to new organs. Moreover, the distribution of radiocesium in the early growth of new organs cannot be clarified by painting leaves with radiocesium. Thus, to determine the translocation to other parts of the trees when the contamination source is assumed to be only the old branches, the branches of grapes from planting areas that experienced the fallout were grafted to uncontaminated trees, and the distribution of ^{137}Cs within the uncontaminated trees was observed at the time of harvest. Hence, the translocation from old organs was investigated.

One-year old branches were sampled from 'Kyoho' grapes planted in an orchard in Fukushima, and were adjusted to a length of 25 cm and 3 buds. Each branch had been grafted to 1-year-old branch parts of 'Kyoho' grapes (4-year-old, 10.0 L potted trees, rootstock cultivar is '101–14') cultivated in a closed greenhouse before the accident. Four trees were selected from those with successful grafts and with confirmed fruiting, and were used as samples in the test (Fig. 11.4). Trees were

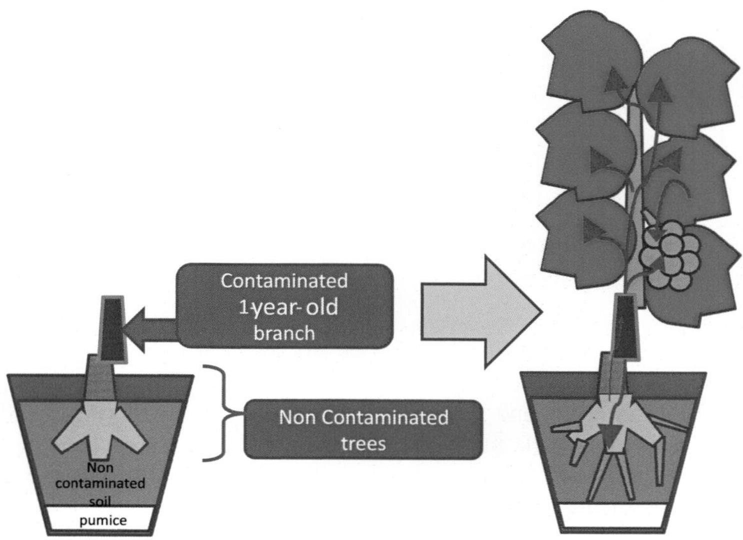

Fig. 11.4 Shifts of ^{137}Cs from scion to other organs in grapes

Fig. 11.5 Photographs of Shifts of ^{137}Cs from scion to other organs in grapes. (**a**) at grafting, (**b**) flowering before 1-week, (**c**) mature fruit (limit the number of flowers, so fruit was smaller than usual)

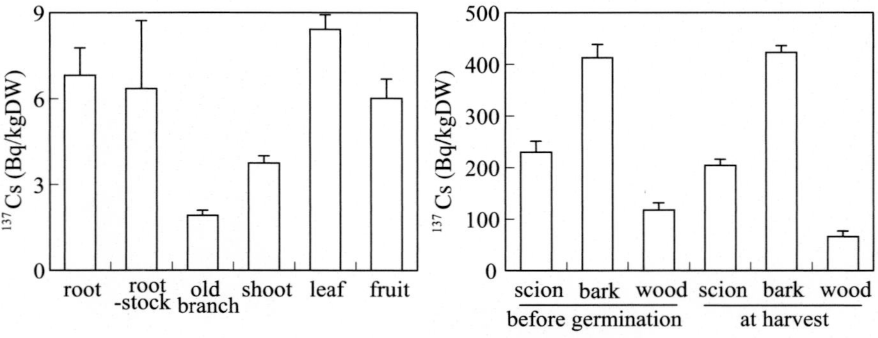

Fig. 11.6 ^{137}Cs concentration of'Kyoho'grapes (from Takata et al. 2013a). *left*: show other than scion at harvest, *right*: scion. Scion homogenized the bark and wood

dissected at harvest time, and ^{137}Cs was measured in each part (Fig. 11.5). The scion was divided into bark and wood, and these were measured. The scion and uncontaminated tree prior to the test were similarly tested for the concentration of ^{137}Cs concentration and content.

At harvest time, among the concentration of ^{137}Cs in each part of a tree, the highest value was found in the scion when the scion was the source of contamination (Fig. 11.6). However, ^{137}Cs was detected from other parts as well. The concentration of ^{137}Cs in new organs positioned above the contaminated scion was the highest in leaves, followed by fruits and then shoots. This was in agreement with the results obtained from grapes cultivated under vinyl covers in the southern region of Fukushima Prefecture (Takata et al. 2012c). It is believed that the concentration of ^{137}Cs tends to be high in leaves. The results obtained for old branches, rootstocks, and roots positioned below the scion showed that the values were similar in roots and rootstocks, and old branches had low values. The hypertrophy of old branches was not observed in this test; therefore, it is thought to play a role in the pathway for translocating ^{137}Cs underground, instead of storing ^{137}Cs, and thus, low ^{137}Cs concentrations were observed in the old branches. In rootstocks and roots, it is possible that accompanying the appearance of pioneer roots and their hypertrophy, that ^{137}Cs stored in new branches was translocated, leading to the similar concentration observed in the fruits. We transplanted contaminated peach trees to uncontaminated soil and investigated the translocation from the old organs (Takata et al. 2012e, 2014b). The concentration of ^{137}Cs in pioneer roots was higher than that in the leaves. Grape roots used in this test were a combination of both the old roots that existed prior to sprouting and the pioneer roots that appeared after sprouting. Due to the browning of the roots, identification was impossible, so all roots were grouped together. However, since concentrations were also high in the pioneer roots in grapes, it is thought that this led to high concentrations in the whole roots.

When the scion was divided into bark and wood, the concentration of ^{137}Cs in the bark did not change much from pre-sprouting to harvest time, but the concentration in the wood decreased to 56.1 % (details are provided in the next section). However, according to a study on the ^{137}Cs concentration in peach bark and wood over time, the ^{137}Cs concentration in the wood temporarily decreased after sprouting, and continued to decrease until harvest (Takata et al. 2013c). Grapes also use stored nutrients to develop new organs immediately after sprouting. Since ^{137}Cs quickly translocated to other organs, similar to other stored nutrients, we believe that the concentration in the wood was low at harvest in this test. In the

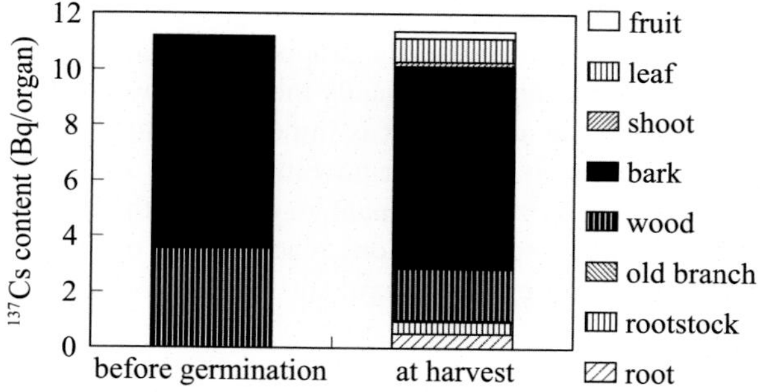

Fig. 11.7 ^{137}Cs content of 'Kyoho' grapes (from Takata et al. 2013c). Content = weight per each organ (gDW) × Bq/kgDW. Bark and wood originated in scion

wood, 19.5 % of ^{137}Cs was redistributed to the other organs. Similar to the concentration results, it was mostly redistributed to leaves, rootstocks, and roots (Fig. 11.7).

11.2.3 Are Transfer Factors in Fruit Trees that Use Soil Concentration as a Guideline Important?

Fruit trees are agricultural products, so radionuclide concentrations in the edible parts are important, and investigating the translocation to the fruits of grapes is fundamentally important. However, transfer factors studied in the existing tests examined translocation from the soil to fruits, so they cannot be directly compared to the proportion of ^{137}Cs in old branches translocating to fruits, as was examined in our current tests. Thus, we attempted to compare translocation from the soil to fruits with the translocation inside the trees to fruits. Transfer factors from the soil varied greatly in grapes depending on the conditions. According to previous reports, the transfer factor was assumed to be 0.00079, a value widely accepted in Japan (MAFF 2011). Let us assume that the ^{137}Cs concentration in the soil was uniformly 10,000 Bq in the planting area. Then, the concentration in fruits was 7.9 Bq/kg FW. If we express this value per dry weight of the fruits in order to compare with our test, we obtain 44.1 Bq/kg DW. Since the ^{137}Cs concentration in the fruits in this test was 6.0 Bq/kg DW, due to the absorption rate from the soil, the ^{137}Cs concentration in the soil must be 1359.5 Bq/kg DW in order to reproduce this value. Furthermore, if we assume that it was planted in the soil with a capacity similar to our test (6.0 kg dry soil), the total ^{137}Cs content in the soil would be approximately 8157.0 Bq equivalent. However, the scion concentration in this test was 230.0 Bq/kgDW, which is only about 1/6 of the calculated required concentration in the soil (1359.5 Bq/kg DW). Furthermore, in contrast to the 11.2 Bq equivalent ^{137}Cs content in the scion, the required ^{137}Cs content in the soil would be 8157.0 Bq equivalent, 728 times the total amount needed in branches. Therefore, compared to the radiocesium in the soil, the radiocesium present in the branches has a much higher contribution to the translocation of radiocesium to the fruits, and therefore, this needs to be carefully monitored in the future.

If the translocation of radiocesium from the bark to the internal parts of trees is reduced, it is possible that the proportion of radiocesium derived from soil detected in fruits and leaves will gradually increase. However, because only a few years have passed since the accident, it is important to focus on the presence of radiocesium inside the trees. It is unclear how many years it takes before the contribution from the soil becomes more dominant, or even whether the condition changes at all. If the changes in fruit concentrations reach a state of equilibrium, it may be used as an index. In any case, a long-term survey is necessary.

11.2.4 Temporal Changes in the Concentration, Especially Translocation from the Bark

In peach trees 4 months after the accident, radiocesium was present at a higher concentration in the outermost layer of the bark than in the soil (Takata et al. 2012a, d); thus, radiocesium absorbed from the aboveground parts of trees was reported to be higher than radiocesium absorbed from the soil through the roots (Takata et al. 2012b). Furthermore, in the previous section, we mentioned that the radiocesium concentration in the bark did not change much, and that the radiocesium concentration in the wood decreased. From these results, understanding the changes in radiocesium in the old branches is as important as finding the transfer factors from the soil in order to sustain future fruit tree production. Therefore, here we present changes observed in radiocesium concentrations in peach bark and wood over time.

Two-year-old branches (3-year-old branches in 2012) from three trees were sampled from 6-year-old 'Campbell Early' peach trees (*Prunus persica* L. Batsch) in a planting area in Tokyo (andosol), every 2–3 months from July 2011 to November 2012. A part of the sampled bark was divided into the outer and inner bark with the cortical layer as the boundary. For each 10 cm length of the branch, radiocesium and ^{40}K contents were determined.

Radiocesium concentrations in the wood did not change in 2011 (Fig. 11.8, right), but decreased from February to May 2012, around the time of blooming. The ^{40}K concentration increased from summer to fall in both years. It is believed that the following factors are related to this trend: (1) with soil as a rich source, ^{40}K is translocated to the aboveground parts through the roots and is then accumulated in the wood, which plays the role of a storage organ; and (2) to enhance this accumulation, fertilizer containing K was applied in August as a topdressing after the harvest. Around the time of blooming in 2012, the ^{40}K concentration decreased

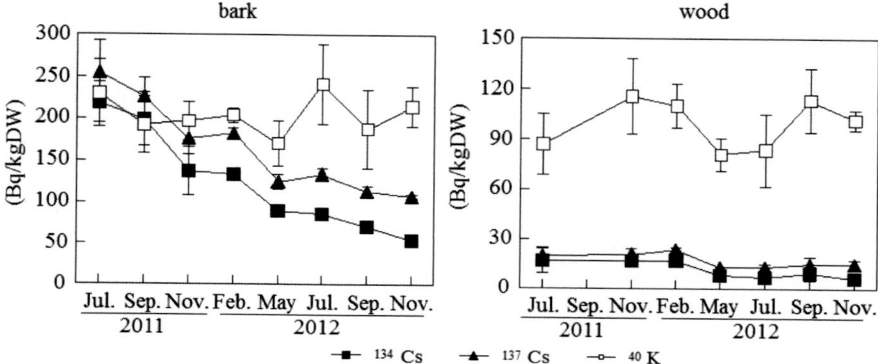

Fig. 11.8 Seasonal change in radiocaesium and ^{40}K concentration of peach lateral shoot (from Takata et al. 2013c). *left*: bark, *right*: wood, n = 3

similar to the radiocesium concentration. Although the concentration of ^{40}K in the bark showed seasonal variation, it changed by roughly 200 Bq/kg DW (Fig. 11.8, left). On the other hand, the radiocesium concentration decreased with time since July 2011 when the survey started. A gradual decrease was noted in ^{137}Cs, which has a long half-life. Contributions from factors other than physical attenuation are likely to be significant. These factors include spontaneous peeling of bark and the translocation of radiocesium in the bark to other organs. We discuss the possibilities of these two factors below.

We can estimate bark peeling as follows: the radiocesium concentration in the bark does not change much during the fruit-growing period and the winter. However, it decreases during autumn when there is active nutritional development. This coincides with the period when parts of the outer bark peel easily as old branches swell. It is possible that the concentration of Cs decreased in association with this phenomenon. To clarify this point, the bark was divided into the outer and inner bark, and ^{137}Cs was measured. The radiocesium concentration in the outer bark decreased from July to November, then November to May, in both years (Fig. 11.9, left). Three factors possibly affecting this trend are: (1) translocation of radiocesium to the inner bark, (2) peeling of parts of the outer bark, and (3) the dilution effect of radiocesium to the outer bark due to the hypertrophy of the cortical layer.

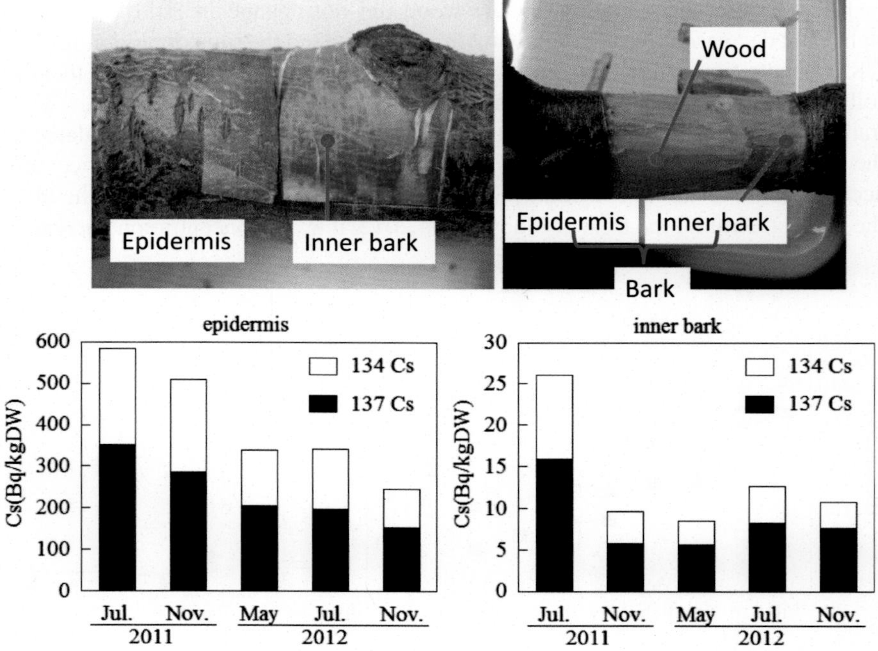

Fig. 11.9 Seasonal change in radiocaesium concentration of peach bark (from Takata et al. 2013c) *left*: epidermis, *right*: inner bark. *: indicate significant difference compare with last date by *t*-test (P = 0.05)

We can assume that if radiocesium in the outer bark is translocated to the inner bark, the concentration within the inner bark and wood would increase; however, radiocesium concentrations in the inner bark and wood did not increase; hence, this possibility is unlikely. Therefore, we can assume that it is due to peeling of parts of the outer bark or dilution due to the development of the cortical layer. However, to quantitatively understand the translocation of radiocesium from the outer bark, it must be examined after determining the level of outer bark peeling and cortical layer development. Thus, it cannot be determined solely based on our results. In addition, the change in the concentration of radiocesium in the whole bark is similar to the change of ^{137}Cs concentration in the outer bark; therefore, it was considered to be easily influenced by the changes in the outer bark at higher concentrations.

The radiocesium concentration in the inner bark decreased significantly from July to November 2011, but showed no change in 2012 (Fig. 11.9, right). In 2011, there was a possibility that the radiocesium in the inner bark translocated to the inner organs (wood), or the outer organs (the outer bark). However, since November 2011, the concentration in the inner bark did not change much; therefore, it is possible that there was little translocation from or towards the inner bark. It is possible that some changes occurred in the permeability within the bark, such as the form of radiocesium, following the accident.

If the decreased concentration of the outer bark was due to the peeling of the bark, peeling organic matter containing a large amount of radiocesium may have re-adhered to new organs, such as leaves and fruits. There are many unknowns about the translocation of radiocesium derived from organic matter, such as bark, to fruits after it re-adheres to leaves and young fruits. However, it is difficult to wash the pericarp surface layer of fruit trees, such as peaches, prior to shipment. Thus, for such fruit trees, manual removal treatments, such as cleaning the bark positioned closest to the fruits and shaving the outer layer of bark, will lower the risk of directly contaminating the surface of fruits through secondary adherence.

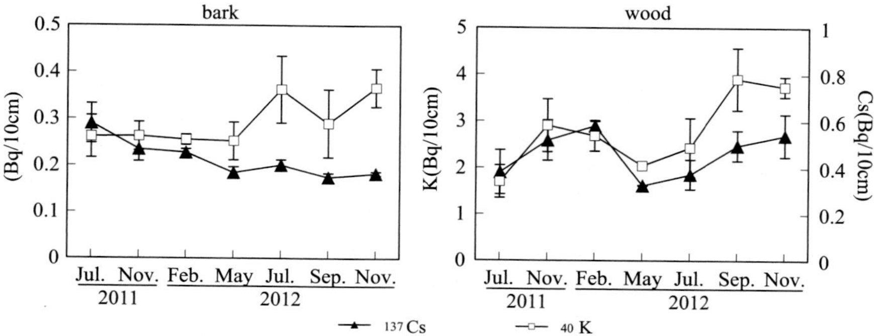

Fig. 11.10 Seasonal change in ^{137}Cs and ^{40}K content of peach lateral shoot (from Takata et al. 2013c) *left*: bark, *right*: wood, (n = 3). Content = weight per 10 cm lateral shoot (kgDW) × Bq/kgDW

The nuclide content per 10 cm of branch was obtained by multiplying the organ concentration by organ weight (Fig. 11.10). The ^{137}Cs content in the bark showed a gradual decreasing trend, similar to the change in ^{137}Cs concentration. The ^{40}K content was also similar to the ^{40}K concentration, but was higher in 2012 than 2011, and showed large variability. This was thought to be due to hypertrophy of the bark and difference in the level of hypertrophy depending on the part. The ^{137}Cs content in the wood was lower than the ^{40}K content, but the manner of change in 2011 was similar to that of ^{40}K. In May 2012, the ^{137}Cs and ^{40}K contents in the wood decreased for a period, and then increased again in the autumn. We believe that the increase in the weight associated with the hypertrophy of the wood played a major role in the increase of the nuclide content of the wood in the autumn in both years. However, the level of increase was different each year, and ^{137}Cs showed a lower value compared to ^{40}K in 2012. This is likely because the portion of radiocesium that decreased in the inner bark in the fall and winter of 2011 was translocated to the wood, while such translocation did not occur in 2012. Thus, the level of increase compared to ^{40}K was different each year. Therefore, in 2012, when there was no translocation of radiocesium from the inner bark, the increased radiocesium in the wood was likely to have derived from the soil. Thus, it is possible that there is a difference in the translocation of ^{137}Cs and ^{40}K to the aboveground parts from the current orchard soil. In particular, it is said that the radiocesium was high in the surface soil layer (5 cm) (Shiozawa et al. 2011), but the roots of peaches are often at 5 cm or deeper, so it is assumed that ^{137}Cs, compared to ^{40}K, was not absorbed as much from the soil.

Based on these results, the radiocesium concentration in the peach bark decreased 2 years after the accident, with a decreasing trend even for ^{137}Cs, which has a long half-life. The main reason for this is the decreasing effect of peeling outer bark where radiocesium had accumulated. In the previous section, we noted that the concentration in the wood significantly decreased, but the concentration in the bark did not change much. In addition, considering that the concentration in the inner bark barely changed 8 months after the accident, as shown in this section, it is possible that the translocation of radiocesium to the inner parts of trees from the outer bark becomes more difficult as time passes since the accident.

11.3 Prediction of Mature Fruit Radiocesium Concentration

In some regions in Japan, fruit shipment was self-regulated even if the concentration was below the regulatory requirement in the years following the accident. Recently, a system that measures the radiocesium concentration in agricultural products was established. For example, rice produced in Fukushima prefecture is shipped after a full inspection is conducted. However, for fruits with a short shelf life, shipping after a full inspection could greatly increase the disposal rate.

Therefore, it is expected that a system will be established in which the radiocesium concentration of the harvested fruits is predicted prior to harvest. In this chapter, we have been discussing the translocation of radionuclides inside trees with a focus on radiocesium. It is still difficult to fully understand the changes in concentrations in the actual fruits only based on the existing results.

When predicting the radiocesium concentration in the fruits, although there is an index such as transfer factors, numerical estimates vary greatly in the literature, even for the same fruit. The transfer factors of fruits may vary depending not only on the soil characteristics but also on the condition of trees, such as their age and degree of pruning. However, the factors that affect the coefficients are still unknown. Furthermore, as discussed in the previous section, the effects of direct infiltration of radiocesium from the fallout on the aboveground parts on the concentration of radiocesium in the fruits are more significant than the effects of absorption from the soil through the roots. The existence of radiocesium that directly infiltrates trees like this makes it difficult to predict the concentration in the harvested fruits based on the radiocesium concentration in the soil. Incidentally, the concentration of radiocesium in peach fruits in the accident year was shown to have decreased as the fruits matured (Sato 2012). If it is true that the concentration in the fruits decreases as the fruits mature, it will be possible to predict the radiocesium concentration in the fruits prior to their harvest.

11.3.1 Fruit and Leaf Radiocesium Concentration

In this section, we study the changes in radiocesium concentration in fruits in the year after the accident during the fruit growth period. We sampled 'Akatsuki' peaches (*Prunus persica* L. Batsch.) planted in a commercial growing field on Ryozen, Date City, Fukushima Prefecture. Over time, we sampled fruits and leaves from 15 days after full bloom to harvest time in 2012. From each tree, 10–50 fruits were sampled, while 50 leaves in mid-position on shoots were randomly sampled from the trees. The concentrations of ^{134}Cs and ^{137}Cs per dry weight of fruits were both at their highest 15 days after full bloom, they decreased from 30 to 50 days after full bloom, and then remained nearly constant until the mature stage (Fig. 11.11). Changes in ^{134}Cs and ^{137}Cs were similar during the fruit growth period, and ^{134}Cs changed to about 70 % of ^{137}Cs. The ^{40}K concentration in the fruit was at its highest 15 days after full bloom, similar to the radiocesium concentration. It decreased after that, but the ratio by which it decreased was different. In other words, although the concentrations of ^{137}Cs and ^{40}K were nearly identical 15 days after full bloom, these values greatly diverged from each other 30 + days after full bloom. The radiocesium concentration in the leaves 15 days after full bloom was higher than that in other periods. On the other hand, the ^{40}K concentration in the leaves increased from 15 to 60 days after full bloom and gradually decreased thereafter. In the existing reports, there are planting areas where the total K concentration in peach leaves increased from 30 to 60 days

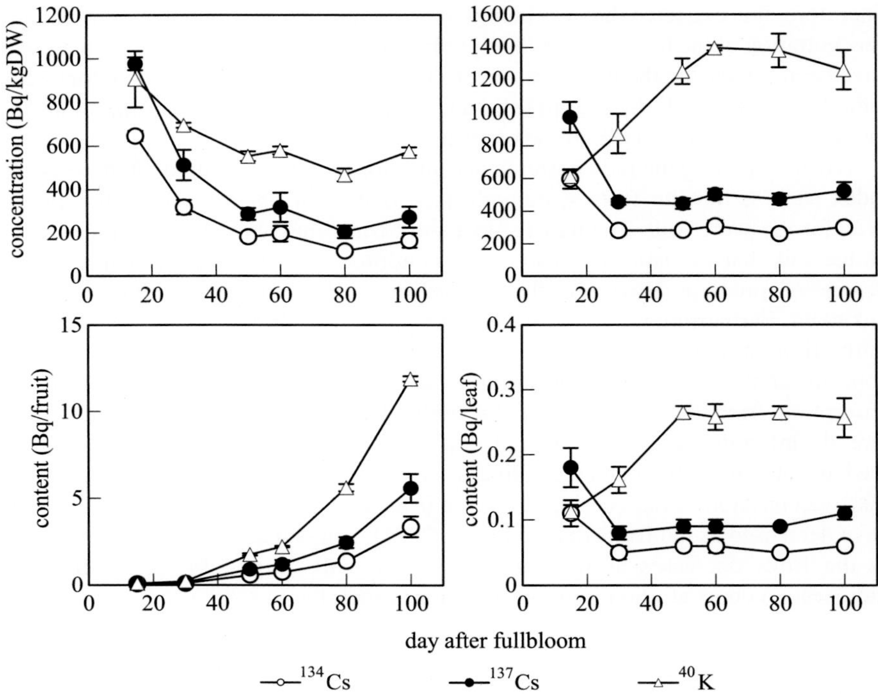

Fig. 11.11 Changes in radiocaesium and ^{40}K concentration (*upper*) and content (*lower*) of peach fruit (*left*) and leaf (*right*) in periods of fruit development (from Takata et al. 2014a, b) *upper*: Bq/kgDW, lower: Bq/one fruit or leaf. n = 3

after full bloom (Takano 2010). It has also been reported that the concentration decreased gradually from 60 to 120 days after full bloom. The ^{40}K in the leaves in our report behaved similarly. Based on these findings, it became clear that there were times when the translocation of radiocesium and ^{40}K to fruits and leaves differed. The radiocesium concentration in the fruits cannot be estimated solely from a simple ratio calculation (Cs/K ratio) based on the K concentration. In addition, we conducted similar surveys in the same planting area in 2013 and 2014. The changes in the radiocesium concentration during the same period showed similar results (unpublished data). On the other hand, the changes in the K concentration showed significant variation depending on the year, and the ^{137}Cs concentration in the harvested fruits showed a 1/3 to 1/4-fold decrease after about 1 year.

11.3.2 Comparison of the Mature Fruit and Thinning Fruit

Compared to radiocesium, ^{40}K sometimes showed a different modality in the leaves; hence using these values to predict the concentration in harvested fruits is difficult. On the other hand, since the year of the accident, the radiocesium concentration in the peach fruits decreased with maturity. Therefore, from each orchard, we sampled 3 'Akatsuki' trees (from 5- to 24-years old) from each of the 24 peach orchards in Fukushima Prefecture at 60 days after full bloom (which is the thinning time) and at maturity. Then we examined the possibility of predicting the radiocesium concentration in the mature fruits using thinned fruits. In Japanese peach production, the amount of fruits is adjusted through steps such as disbudding, deblossoming, and primary and secondary thinning. Sampling fruits during these periods may be less labor-intensive than separately sampling fruits. Therefore, for the sampling period, we chose 60 days after the full bloom as the hypertrophy of fruits would be in progress and the Bq per fruit starts to increase (Fig. 11.12). Additionally, consideration was given in this test to the moisture condition of peach

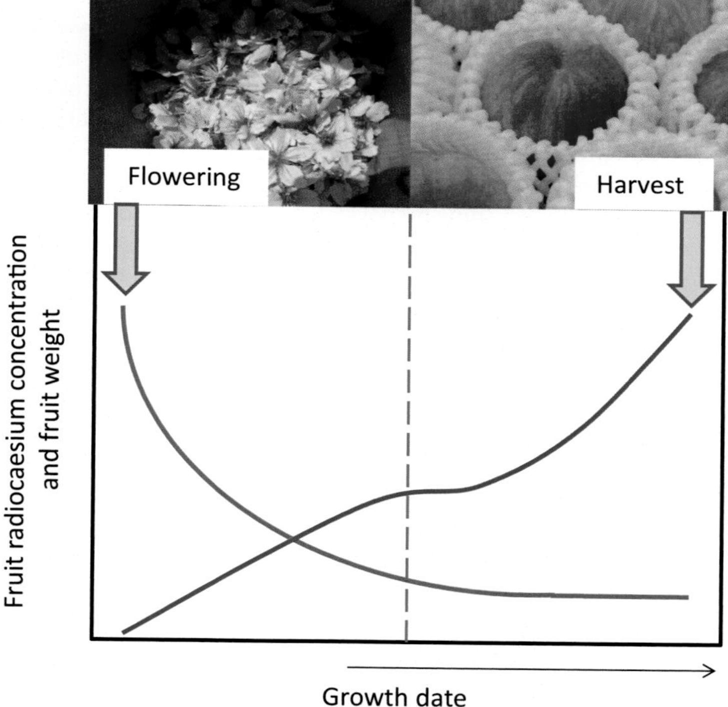

Fig. 11.12 Target timing of prediction. *Red curve*: peach fruit radiocaesium concentration, *blue curve*: peach fruit weight, *green line*: target timing of prediction (using thinning fruits)

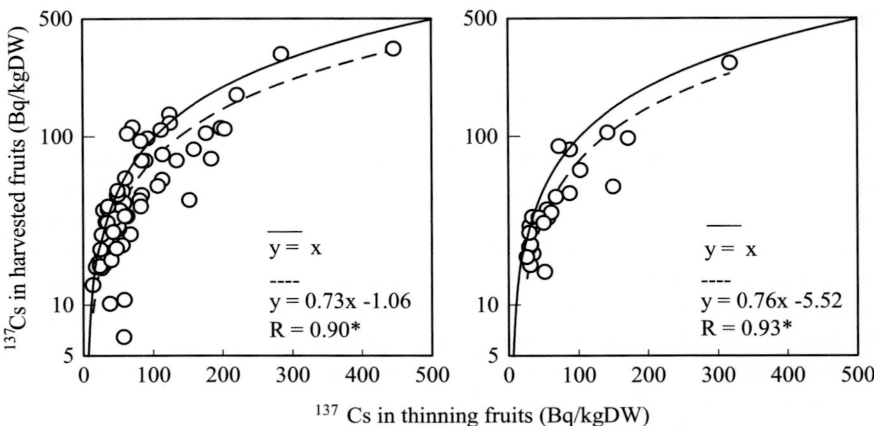

Fig. 11.13 Relationship between thinning fruit and harvested fruits in ^{137}Cs (from Takata et al. 2014a, b) *left*: each tree, (n = 70), *right*: each orchard, (n = 24). *Dashed line* means correlation line. *Solid line* means Y = X line

fruits changing with the weather, conditions of the planting area, and days passed since sampling. The radiocesium concentration was expressed on a dry weight basis (Bq/kg DW). Figure 11.13 shows the relationship of ^{137}Cs between the harvested fruits and fruits sampled at the time of thinning 60 days after full bloom for all 72 trees. When ^{137}Cs concentrations of thinned and matured fruits were compared, a significant and strong correlation was noted. Based on that, it is possible to estimate the concentration in the harvested fruits by measuring the radiocesium concentrations in the thinned fruits 60 days after full bloom. However, to predict the safety of the harvested fruits by actually using the thinned fruits, it is more important to know how many trees exist for which the concentration in the harvested fruits is higher than in the thinned fruits, than it is to calculate the predicted radiocesium concentration of the harvested fruits by substituting values into the obtained correlation equation. Therefore, we added a solid line showing the Y = X relationship. Values plotted above and to the left of this solid line indicate that the ^{137}Cs concentration is higher in the harvested fruits than in the thinned fruits. Thus, 9 out of 70 trees were "outliers" that had higher values in 60th-day thinned fruits than in the matured fruits. To maintain safety, it is important to eliminate such outliers. As a countermeasure, using a method to accelerate the measurement period of the thinned fruits may be an option. As mentioned earlier, the radiocesium concentration is higher in fruits 30 days after full bloom than 60 days after full bloom. Thus, it may be that the accuracy of the safety prediction for harvested plants may be increased. However, in this case, the ^{137}Cs concentration was higher than that in our samples from 60 days after full bloom, although the radiocesium content per fruit was low. Therefore, it is important to note that the number of sampled fruits at the time of measurement was quite high, and that the change in concentrations in the fruits should be considered because it is a period when the variations in temperature and planting area affect the conditions of fruit growth. On one hand, the concentration of the harvested fruits cannot be the same or above that of the concentration 30 days after full bloom, yet there is a possibility of overestimating the concentration of the harvested fruits. Thus, for trees that surpassed the regulatory required value at 30 days after full bloom, we must consider the necessity of retesting at 60 days or later. The second possible countermeasure is a method using safety coefficients that produces higher safety estimates by adding a coefficient to values such as calibration curves. For example, when setting $Y = X + 10$ without changing the slope of the curve, 68 out of 70 trees were at or below the level of thinned fruits + 10.0 Bq/kg DW, so the number of outliers was reduced. Although we must carefully examine the value added as a safety coefficient, it is very unlikely that the fruits from the trees that had sufficiently low radiocesium concentrations in the thinned fruits would suddenly present extremely high values at the time of harvest; thus, it is worth considering. Furthermore, by measuring the radiocesium concentration of the thinned fruits ahead of time, compared to measuring everything during a short harvest period, less concentrated labor periods can be achieved. Additionally, if one attempts to use the obtained values for the following year by observing the change in one tree over time, it is unlikely to increase compared to the previous year; therefore, it might be possible to exclude trees with low concentrations from the measurement target in the following year.

Next, we investigated the correlation for the same set of data for each of 24 planting areas. Out of 24 planting areas, 23 had lower ^{137}Cs concentrations in the mature fruits than the thinned fruits. Tanoi et al. (2013) showed that individual variations exist in rice in a paddy field, so it is quite possible that the radiocesium concentrations would vary in each tree in an orchard. Therefore, it is premature to determine the concentration of the whole planting area based on several trees. However, at least planting areas with high radiocesium concentrations can be chosen, which can possible reduce the effort required to measure individual trees later on. Of course, it is possible to determine the safety of trees by tests conducted during this period, excluding the trees in planting areas with values close to the shipment limits for radiocesium.

11.4 Conclusions

Since radiocesium is distributed on the surface layer of the soil in orchards, it is important to focus on the translocation of radiocesium from the soil to trees when they have shallow roots. Moreover, understanding the behavior of radiocesium in old organs is equally significant. If the concentrations in the bark and wood are obtained separately, a better understanding of the proportion of radiocesium translocated to the fruits is possible, because the translocation of radiocesium from the bark to the internal part of trees becomes more difficult as time passes after an accident.

We confirmed that there was a difference in the behaviors of radiocesium concentration and ^{40}K concentration in the fruits and leaves of peaches. Furthermore, differences can arise between the behaviors of radiocesium and K inside the trees; therefore, it is risky to consider the relationship of radiocesium and K in perennial crops as similar to single annual crops that have only soil as a radiocesium source.

Comparison of the concentration of ^{137}Cs in the fruits sampled 60 days after full bloom with that of harvested fruits revealed a strong correlation. However, the concentrations in the harvested fruits were not always lower than the fruits sampled 60 days after full bloom. Since trends in the planting area and the internal parts of trees can be understood to a certain degree by measuring the thinned fruits at 60 days, this indicates that the safety could be determined through tests during that time, excluding planting areas and internal parts of trees with numerical values extremely close to the shipment limit.

Acknowledgments This article is based on the collaborative research with Dr. Eriko Yasunaga, Dr. Haruto Sasaki, Dr. Keitaro Tanoi, Dr. Natsuko Kobayashi, Dr. Seiichi Oshita (the University of Tokyo) and Mamoru Sato, Kazuhiro Abe (Fukushima Prefecture Agricultural Technology Centre), with the assistance of Mr. Kengo Izumi, Mr. Kenichiro Ichikawa (the University of Tokyo). This study was partly supported by of Ministry of agriculture, forestry and fisheries (MAFF, Japan).

References

Carini F, Scotti IA, D'Alessandro PG (1999) ^{134}Cs and ^{85}Sr in fruit plants following wet aerial deposition. Health Phys 77:520–529

IAEA (2003) Report of the fruits working group. Modeling the transfer of radionuclides to fruit, pp 12–21

Marouf BA, Mohamad AS, Taha JS, Tawfic NF, Mahmood JA (1992) The transfer of ^{137}Cs from soil to plants. Environ Int 18:211–215

Ministry of Agriculture, Forestry and Fisheries, Japan (2011) TF of vegetables and fruits in the farmland soil. http://www.maff.go.jp/j/press/syouan/nouan/pdf/110527-01.pdf. Accessed 26 June 2007

Ministry of Agriculture, Forestry and Fisheries (Japan) (2013) Statistical yearbook of ministry of agriculture, forestry and fisheries. http://www.maff.go.jp/j/tokei/kouhyou/sakumotu/menseki/#r, http://www.maff.go.jp/j/tokei/kikaku/book/seisan/25_kaju.html

Oshita S, Yasunaga E, Takata D, Tanoi K, Kawagoe Y, Shirai Y, Nakanishi TM, Sasaki H, Makino Y (2013) Radioactivity measurement of soil and vegetables arisen from Fukushima Daiichi nuclear power plant accident: case studies of low level contamination in Tokyo and Fukushima. Radioisotopes 62:149–157 (in Japanese with English abstract and tables)

Sato M (2012) Fruit in Fukushima in the nuclear accident year: report of the first year examination to take measures against the radioactive contamination of nuclear power plant accident. Radiochem News 26:21–31 (In Japanese)

Shiozawa S, Tanoi K, Nemoto K, Yoshida S, Nishida K, Hashimoto K, Sakurai K, Nakanishi TM, Nihei N, Ono Y (2011) Vertical concentration profiles of radioactive caesium and convective velocity in soil in a paddy field in Fukushima. Radioisotopes 60:323–328 (in Japanese with English abstract and Tables)

Takano K (2010) Delicious peach fruit production and shipping technology. Bull Okayama Agric Res 1:23–90

Takata D (2013) Distribution of radiocaesium from the radioactive fallout in fruit trees. In: Nakanishi TM, Tanoi K (eds) Agricultural implications of the Fukushima nuclear accident. Springer, Japan, pp 143–162

Takata D, Yasunaga E, Tanoi K, Nakanishi T, Sasaki H, Oshita S (2012a) Radioactivity distribution of the fruit trees ascribable to radioactive fall out: a study on stone fruits cultivated in low level radioactivity region. Radioisotopes 61:321–326 (in Japanese with English abstract and tables)

Takata D, Yasunaga E, Tanoi K, Nakanishi T, Sasaki H, Oshita S (2012b) Radioactivity distribution of the fruit trees ascribable to radioactive fall out (II): transfer of raiocaesium from soil in 2011 when Fukushima Daiichi nuclear power plant accident happened. Radioisotopes 61:517–521 (in Japanese with English abstract and tables)

Takata D, Yasunaga E, Tanoi K, Kobayashi N, Nakanishi T, Sasaki H, Oshita S (2012c) Radioactivity distribution of the fruit trees ascribable to radioactive fall out (III): a study on peach and grape cultivated in south Fukushima. Radioisotopes 61:601–606 (in Japanese with English abstract and tables)

Takata D, Yasunaga E, Tanoi K, Nakanishi T, Sasaki H, Oshita S (2012d) Radioactivity distribution of the fruit trees ascribable to radioactive fall out (IV): cesium content and its distribution in peach trees. Radioisotopes 61:607–612 (in Japanese with English abstract and tables)

Takata D, Sato M, Abe K, Tanoi K, Kobayashi N, Yasunaga E, Sasaki H, Nakanishi T, Oshita S (2012e) Remobilization of radiocaesium derived from Fukushima nuclear power plant accident in the following year in 'Akatsuki' peach trees. Hort Res (Japan) 11(suppl 2):353 (in Japanese)

Takata D, Sato M, Abe K, Yasunaga E, Tanoi K (2013a) Radioactivity distribution of the fruit trees ascribable to radioactive fall out (V): shifts of caesium-137 from scion to other organs in 'Kyoho' grapes. Radioisotopes 61:455–459 (in Japanese with English abstract and tables)

Takata D, Sato M, Abe K, Yasunaga E, Tanoi K (2013b) Radioactivity distribution of the fruit trees ascribable to radioactive fall out (VI): effect of heterogeneity of caesium-137 concentration in soil on transferability to grapes and fig trees. Radioisotopes 62:533–538 (in Japanese with English abstract and tables)

Takata D, Yasunaga E, Tanoi K, Nakanishi T, Sasaki H, Oshita S (2013c) Radioactivity distribution of the fruit trees ascribable to radioactive fall out (VII): seasonal changes in radioceasium of leaf, fruit and lateral branch in peach trees. Radioisotopes 62:539–544 (in Japanese with English abstract and tables)

Takata D, Sato M, Abe K, Kobayashi N, Tanoi K, Yasunaga E (2014a) Radioactivity distribution of the fruit trees ascribable to radioactive fall out (VIII): prediction of peach fruit radiocaesium concentration by thinning fruits. Radioisotopes 63:293–298 (in Japanese with English abstract and tables)

Takata D, Sato M, Abe K, Tanoi K, Kobayashi N, Yasunaga E (2014b) Shift of radiocaesium derived from Fukushima Daiichi nuclear power plant accident in the following year in peach trees. In: 29th international horticultural congress, impact of Asia-Pacific horticulture 117, p 215

Tanoi K, Kobayashi N, Ono Y, Fujimura K, Nakanishi TM, Nemoto K (2013) Radiocaesium distribution in rice plants grown in the contaminated soil in Fukushima prefecture in 2011. Radioisotopes 62:25–29

Zehnder H, Kopp JP, Eikenberg J, Feller U, Oertli JJ (1995) Uptake and transport of radioactive cesium and strontium into grapevines after leaf contamination. Radiat Phys Chem 46:61–69

Chapter 12
The Effects of Radioactive Contamination on the Forestry Industry and Commercial Mushroom-Log Production in Fukushima, Japan

Satoru Miura

Abstract The accident at the Fukushima Daiichi nuclear power plant in 2011 left surrounding residential, agricultural, and forested areas contaminated with radiation on a massive scale. To encourage evacuees to return to their homes and resume agricultural practices, large-scale decontamination of radioactive zones is ongoing in residential and agricultural areas. However, contamination of forests is extensive and decontamination efforts have been limited by remote access, significant labor requirements, and the considerable amount of anticipated radioactive waste. Consequently, there has been no large-scale effort to decontaminate forests as there has been for residential and agricultural land. In this paper, we examine the current protection of forests from radioactive contamination and discuss measures required to promote forest restoration. In addition, we consider how forest contamination relates to radiation exposure in humans and summarize the state of the forestry industry since the Fukushima accident. We also consider how radiation affects forest products in Fukushima, emphasizing mushroom-log production. Finally, we examine the challenges surrounding the reconstruction and revival of forests and forestry in Fukushima.

Keywords Forest restoration • Hardwood forest management • Mushroom cultivation • Mushroom logs • Radiocesium transfer

S. Miura (✉)
Department of Forest Site Environment, Forestry and Forest Products Research Institute, Matsunosato 1, Tsukuba, Ibaraki 305-8687, Japan
e-mail: miura@ffpri.affrc.go.jp

T.M. Nakanishi, K. Tanoi (eds.), *Agricultural Implications of the Fukushima Nuclear Accident*, DOI 10.1007/978-4-431-55828-6_12

12.1 Current Forest Contamination and Remediation Efforts

12.1.1 External Exposure

The effect of forests contaminated with radiation on humans depends on whether the radiation exposure is external or internal. External exposure can occur through either close contact with contaminated timber or exposure to contaminated forest sites during outdoor activities. The highest level of radioactive contamination found in timber from Fukushima Prefecture was 414 Bq/kg (479 Bq/kg dry weight) and in 2012, the Forestry Agency, Japan estimated that a person living in a room made from this contaminated timber would incur an additional 0.012 mSv per year (Forestry Agency 2012a). This is 1.2 % of the accepted additional dose of 1 mSv per year and would have a negligible effect on human health. However, in Japan, there is no regulation of the radiation levels of building materials taken from conifer forests, and therefore the Fukushima Lumber Co-operative Union has adopted a limit of 1000 gamma ray counts per minute, using a Geiger–Müller counter, as this is the permissible level for contaminated materials from a laboratory with radioisotope analysis (Fukushima Lumber Cooperative Unions 2012). This self-regulation and monitoring of timber radiation levels is to counter and refute unfounded concerns regarding the use of timber from Fukushima Prefecture and to promote its distribution.

Radiation exposure sustained during outdoor activities in contaminated forests depends on the air dose rate and time spent in contaminated areas. This is of concern to forestry workers and decontamination crews, due to the significant amounts of time they might spend in contaminated forests. Furthermore, contaminated forests will also affect people visiting for recreational activities, although they generally spend less time in contaminated areas than forestry workers. To ensure safer conditions, the government has enforced the Ordinance on Prevention of Ionizing Radiation Hazards, which aims to protect workers by managing exposure doses based on the air dose rate and exposure to radioactive materials in the workplace (Ministry of Health, Labour and Welfare, Japan 2013a). For workers other than those involved in decontamination, this law will keep occupational exposure dosage below the limit set by the International Commission on Radiological Protection (Valentin 2007). In addition, the Forestry Agency and the Fukushima Prefectural government are implementing policies to reduce exposure rates to forestry workers, such as equipping large forestry machinery with lead shielding and using substances to reduce the air dose rate in contaminated areas (Forestry Agency 2014a). Moreover, the Nuclear Regulation Authority is providing safety information and countermeasures to reduce radiation exposure to people returning home (Nuclear Regulation Authority 2013). Finally, government policy has shifted its emphasis from estimates based on air dose rates to values obtained from dosimeters that track individuals.

12.1.2 Internal Exposure

Consumption of contaminated food is the primary means of internal radiation exposure to humans. Forests in Japan produce many food products, including *sansai* (edible wild plants), bamboo shoots, fruits, nuts, and wild game, but mushrooms are by far the most economically important forest food product (Forestry Agency 2014f). Additionally, artificially cultivated mushrooms are the only forest food product that is consumed on a large, commercial scale.

Immediately after the Fukushima accident in March 2011, the Japanese government set a maximum tolerable level of radioactivity in food at 500 Bq/kg for adults to protect against internal radiation exposure (Ministry of Health, Labor and Welfare, Japan 2011). Prefectural governments inspected the radioactivity of foods and reported findings to the Ministry of Health, Labor and Welfare, and the government prevented distribution and shipping of food that exceeded the radiation limits. This monitoring system applied to conventional agriculture as well as food grown in forests. The Fukushima government abided by and implemented such a monitoring system to inspect agricultural products and protect the public from consuming excessively contaminated food. For example, rice, which is the main agricultural product in Fukushima, was inspected bag-by-bag in 2012 to allay consumer fears of contamination (Nihei et al. 2015). The government initiated similar inspections at the shipment stage to ensure the safety of forest food products, including mushrooms. However, in April 2012, the Japanese government revised the radioactive cesium limit for general food to 100 Bq/kg (Ministry of Health, Labour and Welfare, Japan 2012).

The Ministry of Health, Labor and Welfare, Japan set these regulations for food in the general market, but no such guidelines were set for food produced, hunted or gathered by individuals for their own use. This is of particular concern, as people often collect mushrooms and *sansai* in the mountainous areas of Fukushima. Shortly after the Fukushima nuclear accident, the government published information about radioactive contamination, including a radiation map and shipping restrictions on agricultural and forest products. This was the extent of protection against radiation health effects for people collecting their own food. However, local authorities have provided community centers with the equipment needed to test radiation levels in food to provide comprehensive information regarding the impact of radiation.

The Japanese government monitored radioactive Cs (Cs-134, Cs-137) levels in food through market-basket samples in autumn 2012 at 15 locations nationwide, including Fukushima. All calculated annual doses were less than 1 % of 1 mSv, far below the permissible annual dose (Ministry of Health, Labor and Welfare, Japan 2013b). However, radioactive contamination of forests has affected people economically and altered rural lifestyles, as forests comprise 71 % of Fukushima Prefecture and many people make a living from harvesting forest products. Large-scale radioactive contamination of forests is the main problem interfering with revitalization and reconstruction following the Fukushima nuclear accident.

12.1.3 Ecological Damage and Extent of Radioactive Contamination

In addition to human exposure to radiation through contaminated forests and forest products, radiation has caused ecological loss and damage. Although reports of mutations and ecological devastation are limited, radiation has affected lycaenid butterflies (Hiyama et al. 2012), birds and animals (Ishida 2013), and earthworms (Hasegawa et al. 2013). Moreover, organisms have absorbed radioactive particles from the environment (Murakami et al. 2014). This differs from the Chernobyl nuclear disaster, after which radiation emissions devastated trees and forests over a large area (Arkhipov et al. 1994).

In addition to damage to forests themselves, there is concern that radioactive materials can move from forests into rivers, agricultural land, coastal waters or residential areas. Fallen leaves blowing from contaminated forests to decontaminated residential areas, and irrigation water from streams originating in contaminated forests, could spread radioactivity. The government and universities monitor contamination of forest streams and have found that such contamination is restricted to periods of heavy rain (Forestry and Forest Products Research Institute 2012). When there is no rainfall, streams are not turbid and radioactive Cs are nearly undetectable. However, during heavy rainfall, tumultuous water stirs up radioactive sediment. Still, when suspended matter is filtered out, very little radioactive Cs remain. This indicates that ionic radioactive Cs levels are negligible and that clay particles absorb the majority of the radiation. Furthermore, the amount of radioactive materials flowing in forest streams was found to be less than 1 % of the total deposited material (Hayashi 2013; Takahashi 2013). Conversely, after removing contaminated leaves and other detritus from the forest floor, thereby leaving the ground exposed, considerable amounts of radioactive soil and particulate matter were discharged from the bare ground (Forestry Agency 2014a). This suggests that minimizing soil movement by leaving the forest floor undisturbed and covered with leaves and other organic material will mitigate transfer of radioactive matter from forests to neighboring areas.

Studies indicate that forests act as reservoirs for radioactive material. Fixed-site monitoring in forest ecosystems showed that deposition of radioactive Cs did not change considerably from 2011 to 2013 (Forestry Agency 2014b). Following the Fukushima nuclear accident, there were concerns that radioactive material might spread through soil erosion in steep environments with high rainfall, characteristic of forests in Fukushima. However, even under such environmental conditions, through management of forests and maintenance of forest floor cover, forests will retain the majority of the fallen radioactive material.

Compared to regions contaminated by the Chernobyl nuclear accident, areas around Fukushima are densely populated and forestry activities are more common. However, using forests as a sink for radioactive substances could be a viable option. Decontamination requires collecting contaminants and although it is possible to manage radioactive substances, it is not possible to eliminate radioactivity.

Decontamination strategies should prioritize low-cost management to ensure that radioactive substances do not spread until their radioactivity reaches a safe level.

12.1.4 Summary of Radiation Mitigation in Forests

Most radioactive materials in a forest remain within the forest ecosystem. While there has been no active decontamination program for forests as there has been in agricultural and residential areas, we have pursued means to protect human health. For example, we understand the distribution of radioactive substances in forests and have investigated measures to keep radiation at safe levels for people. Moreover, we have examined how to prevent radioactive contaminants in streams from flowing out of forests. Still, we have not addressed all issues concerning human and environmental health, such as radiation in logs for mushroom cultivation.

Within the context of tree physiology, the problem of radiation in logs begins with the movement of radioactive Cs into trees. After the Chernobyl nuclear disaster, the International Atomic Energy Agency (2002) developed models to predict the movement of radioactive substances into trees and these have since been applied in Japan (Hashimoto et al. 2013). However, data were collected only shortly after the Fukushima accident and additional data are required to improve predictions. Predictions should also be species-specific, especially for species used for cultivation of mushrooms, such as konara oak (*Quercus serrata*). The remainder of this report describes the movement of radioactive cesium into broad-leaf trees used for mushroom cultivation.

12.2 Radioactive Contamination and Mushroom Cultivation

12.2.1 Mushroom Cultivation in Japan

In Japan, mushrooms are a commercially important commodity that are usually artificially cultivated. There are two common methods of cultivating mushrooms: log cultivation (Fig. 12.1a) and sawdust substrate cultivation (Fig. 12.1b). For log cultivation, the fungal inoculum is implanted in small holes drilled in broadleaf logs kept either outside on bare ground, or inside a specialized facility. Sawdust substrate cultivation uses a mix of broadleaf sawdust and rice bran as the growth medium and the mushrooms develop inside specialized facilities. The flavor of log-cultivated shiitake mushrooms (*Lentinus edodes*) is generally preferred to those cultivated using the sawdust method. However, the industry uses the sawdust method more commonly; it accounts for 85 % of shiitake production (Forestry Agency 2014c).

Fig. 12.1 Shiitake mushroom (*Lentinus edodes*) cultures. (**a**) Mushroom-log cultivation; (**b**) Sawdust substrate cultivation (Photos by Hitoshi Neda)

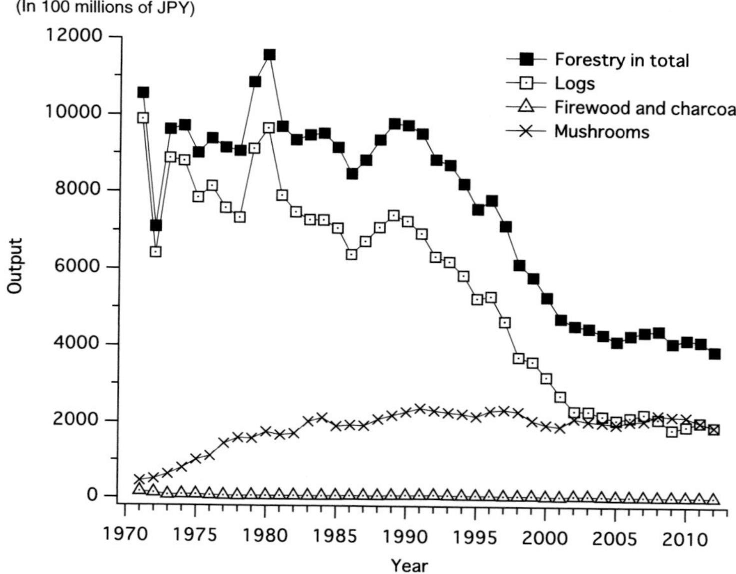

Fig. 12.2 Economic forestry output in Japan from 1971 to 2012

Mushroom cultivation increased gradually through the 1970s and 1980s, while timber production dramatically declined during the 1990s, resulting in the two industries having similar market values by the early 2000s (Fig. 12.2). In 2010, mushroom cultivation was worth 218.9 billion yen, representing 52 % of the value of all forestry products (Table 12.1). However, agriculture and forestry production in Fukushima both dropped sharply after the nuclear accident, and while agricultural revenue was recovering by 2012, forestry revenue, including mushroom cultivation, took one more year to start to recover.

Mushroom cultivation depends on wood as the culture medium. Before the nuclear accident, Fukushima Prefecture was the largest producer of wood for this purpose. In 2009, 10 % of wood for mushroom cultivation was imported from outside the prefecture in which the mushrooms were produced, and Fukushima supplied most of this wood to 22 of 47 prefectures (Forestry Agency 2014d). Hardwood production numbers reflect this, and in 2010 Fukushima Prefecture ranked number three among prefectures in hardwood production by volume in Japan (Table 12.2). After the nuclear accident in 2011, wood supply for mushroom cultivation from Fukushima almost completely ceased, thus affecting mushroom cultivation nationwide. The Forestry Agency and stakeholders in the mushroom industry demanded an adjustment scheme in autumn 2011 (Forestry Agency 2011). However, as of 2014, the mismatch between supply and demand had not been rectified.

Table 12.1 Total agricultural and forestry output and the breakdown of forestry production in Japan and Fukushima before and after the Fukushima nuclear power plant accident

(In 100 millions of JPY)

	Year	Agricultural output	Forestry output	Logs output	Cultured mushrooms	Fuelwood and charcoal
Japan	2009	81,902	4122	1861	2200	49
	2010	81,214	4217	1946	2189	51
	2011	82,463	4166	2055	2047	51
	2012	85,251	3917	1933	1932	44
	2013	84,668	4322	2221	2035	55
Fukushima	2009	2450	130	83	46	2
	2010	2330	125	73	49	2
	2011	1851	87	62	24	1
	2012	2021	74	56	17	1
	2013	2049	86	62	23	1

Data obtained from the Forestry Agency (*Source*: Statistics Department, Minister's Secretariat, Ministry of Agriculture, Forestry and Fisheries)

Table 12.2 Prefectures with the highest hardwood forest production in Japan in 2010 and 2012

(In 10 millions of JPY)

Year	2010		2012	
Rank	Prefecture	Hardwood production	Prefecture	Hardwood production
1	Hokkaido	753	Hokkaido	649
2	Iwate	358	Iwate	346
3	**Fukushima**	152	Kagoshima	144
4	Kagoshima	141	Hiroshima	116
5	Miyazaki	95	Shimane	81
6	Hiroshima	86	**Fukushima**	78
7	Aomori	63	Miyazaki	59
8	Shimane	59	Aomori	52
9	Miyagi	53	Akita	49
10	Akita	53	Yamagata	48
	Japan	2376	Japan	2062

Data obtained from the Forestry Agency (*Source*: Statistics Department, Minister's Secretariat, Ministry of Agriculture, Forestry and Fisheries)

12.2.2 Contamination of Mushroom Growth Media

Mushrooms readily absorb and accumulate Cs (Kalač 2001). After the Chernobyl nuclear accident, mushrooms were among the most contaminated forest products, along with berries and game meat (International Atomic Energy Agency 2006). Following the Fukushima nuclear accident, food inspections across Japan revealed that wild-picked mushrooms had particularly high radioactive contamination; these inspections detected levels above allowable limits in prefectures as distant as Aomori and Nagano. As of autumn 2014, restrictions on mushroom shipments from ten prefectures remained in effect (Forestry Agency 2014e) and contamination of mushrooms was ubiquitous in eastern Japan. However, radioactive emissions from the Fukushima accident had a ^{134}Cs:^{137}Cs ratio of approximately 1:1, which had become 1:3 by autumn 2014 due to their different decay rates. However, most of the radioactive Cs in prefectures far from Fukushima is ^{137}Cs; often no ^{134}Cs is present (Yamada 2013). In such samples, the ^{137}Cs contamination of mushrooms likely originates from atmospheric nuclear testing in the 1950s and 1960s. The Nuclear Regulation Authority has monitored this radioactive fallout since the late 1950s (Nuclear Regulation Authority 2014). Moreover, radioactive ^{137}Cs also reached Japan from the Chernobyl nuclear accident, but declined immediately to a steady level from an initial spike.

In 2011, the Japanese government set acceptable radiation levels for food, including mushrooms, at 500 Bq/kg, which was reduced to 100 Bq/kg in 2012. In response to this, the Forestry Agency immediately started to investigate maximum radioactivity levels for mushroom media to produce mushrooms below the allowable radiation limit. This investigation determined that the maximum contamination value for both log and sawdust media was 150 Bq/kg. However, a follow-up study adjusted these limits to 50 Bq/kg for logs and 200 Bq/kg for sawdust (Forestry Agency 2012b). The lower radioactivity limit for logs is due to greater variation in this medium, although purveyors of these logs consider the limit excessive.

In 2014, a study in Nakadori, Fukushima, found ^{137}Cs levels from 500 to 800 Bq/kg in konara oak logs used for mushroom cultivation (Table 12.3); even the lowest contamination levels were tenfold the allowable limit. Moreover, the study site had a radiation contamination level of 100–300 kBq/m^2 and the surveyed logs tested would have been live trees at the time of the Fukushima nuclear accident. Regardless, it seems that no konara oak logs from Nakadori or Hamadori regions are currently suitable for mushroom cultivation. Conversely, parts of the Aizu region in western Fukushima have lower contamination levels, and subsequently minimal production of logs for mushroom cultivation has been resumed.

The Forestry Agency published guidelines to manage log cultivation and maintain radioactivity within the set limits (Forestry Agency 2013). These include washing the mushroom logs and measures to prevent them from contacting contaminated soils. Another experiment, immersing logs in Prussian blue solution, which adsorbs radioactive cesium and prevents it from entering mushrooms, was extremely effective (Neda 2013). However, this adds to production costs and may color mushrooms blue, thereby reducing their value, rendering the technique unviable. While producing safe mushrooms is the priority, this has negatively affected the outlook for producers of mushroom logs in Fukushima Prefecture.

Table 12.3 Radiocesium concentrations in stem wood and bark, and their weighted averages in logs used for mushroom cultivation[a]

Part of stem	Size	DBH (cm)	^{134}Cs[b](Bq kg^{-1})	^{137}Cs[b](Bq kg^{-1})
Stem wood	Large	12.1, 13.2, 13.6	95 ± 5	271 ± 15
	Medium	9.9, 10.9	93 ± 15	264 ± 38
	Small	7.8, 9.2, 10.0	55 ± 4	159 ± 14
Bark	Large		727 ± 21	2083 ± 71
	Medium		940 ± 85	2657 ± 223
	Small		1047 ± 190	2998 ± 547
Weighted average of stem wood and bark	Large		184 ± 7	524 ± 19
	Medium		279 ± 15	789 ± 42
	Small		250 ± 69	716 ± 199

[a]Miura et al., unpublished data. Field survey was conducted in the Miyakoji area, Tamura, Fukushima in March 2014 (deposited amount of ^{134}Cs and ^{137}Cs by airborne monitoring on July 2 2011 was 100–300 kBq m^{-2}. Air dose rate was 0.33 μSv h^{-1} on March 18 2014)
[b]Average of two or three stems of each individual sample, calculated as the average of three discs at heights of 0.5, 2, and 4 m. Radioactivity of stem wood and bark was determined by a germanium detector with counting error of 3 sigma

12.3 Restoring Hardwood Production

12.3.1 Mushroom Log Production by Forest Cooperatives

Fukushima Chuo Forest Cooperative is overseeing a major mushroom log production operation (Fig. 12.3) in the Abukuma Mountains west of the Fukushima Daiichi nuclear power plant. In 2010, the cooperative had a membership of 8590 and 51,531 ha of private forest. Before the Fukushima nuclear accident, the cooperative successfully focused on producing high-quality hardwood logs for mushroom cultivation and consistently turned a considerable profit. However, now the cooperative must wait to see how long radioactive contamination will impede its hardwood production business.

Commercially grown konara oak and other broadleaf trees regenerate by harvesting trees under 20 years old, as new sprouts stem from the stumps. This coppicing method produces up to three main stems from each stump (Fig. 12.4); the stem's straight portions are used for mushroom cultivation, while the curved or gnarled portions are pulverized for sawdust production. The market price of mushroom logs is currently high, and is two- to fourfold the price of softwood products, such as cedar timber and wood chips for paper production (Hayajiri 2013). Consequently, mushroom log production is a valuable aspect of forestry, although after the Fukushima nuclear accident, this is not the case for the Fukushima Chuo Forest Cooperative.

Before the Fukushima nuclear accident, the Fukushima Chuo Forest Cooperative produced mushroom logs within 20 km of the nuclear plant. After the accident,

Fig. 12.3 A coppiced forest at Miyakoji, Tamura, Fukushima. Two-year-old coppices are in the foreground; an oak forest that has grown beyond its optimal harvesting date due to the Fukushima nuclear accident is in the background

these areas were under evacuation orders until April 2014, at which point forestry activities could resume. Currently, the cooperative's primary concern is whether radioactive contamination levels of konara oak planted now will be below the allowable limit when harvested in 20 years. Trees contaminated after the accident undoubtedly lost their value, but predicted effects on mushroom logs harvested in 20 years will affect forest management. If prospects for future harvests are not favorable, then a major reappraisal of forestry management and production is required. However, if it is possible to produce mushroom logs below allowable radiation contamination levels, then current forest management practices will suffice. Unfortunately, we cannot predict radiation levels of konara oak grown under current conditions.

12.3.2 Resuming Mushroom Log Production

Timber becomes radioactive when radioactive particles migrate into its tissues from the bark or through the roots, and predicting the amount of radioactive cesium that will accumulate in konara oak trunks over the next 20 years is necessary for the mushroom industry. The half-life of ^{134}Cs is 2 years, so in 20 years, it will decrease to 1/1000 of the present level, meaning that only ^{137}Cs is of concern. However, currently, even this isotope alone is above total allowable radiation limits for mushroom cultivation media.

Fig. 12.4 Coppicing regeneration of *Quercus* species (**a**) Mature tree; (**b**) New planting; (**c**) Coppicing

In regenerating coppices, ^{137}Cs enters from the belowground stump and roots, while in newly planted trees, the only source is through root uptake. Therefore, contamination of newly planted trees depends on soil contamination and the tree's ability to absorb contaminants (transfer factors). Studies near the Chernobyl accident have found that transfer factors vary from 10 to 100-fold and depend on soil contamination levels and characteristics, and tree species and age (Shcheglov et al. 2001). However, data regarding how tree species and soil type can influence ^{137}Cs uptake in Japanese hardwood trees are insufficient.

Moreover, radiation dynamics in trees regenerating from stumps are more complex than in newly planted trees. For instance, underground portions of the stump have also been subject to considerable contamination, which would affect contamination of the coppices. However, there is limited information concerning root contamination. In addition, although coppices grow more rapidly than newly planted trees because of the considerable carbohydrate stores in the stump, ^{137}Cs migration rates into the coppice are unknown. Migration rates differ greatly depending on the growth characteristics of the tree. If Cs migrates to the roots and inhibits uptake of new cesium, then tree growth over 20 years would dilute the Cs in above- and below-ground tissues. When considered with radioactive decay, contamination in new coppices may be considerably lower than in trees directly contaminated from the Fukushima accident. However, present knowledge regarding radioactive contamination in hardwood trees is insufficient to confirm this speculation.

To predict ^{137}Cs contamination behavior in coppices, further research into trees' absorption of Cs, its migration through the bark into the tree, and its movement and distribution in the tree is required. Twelve years after the Chernobyl accident, ratios of radioactive ^{137}Cs and stable ^{133}Cs were constant throughout pine and birch stands (Yoshida et al. 2004). This shows that the Cs concentrations in forest ecosystems reached an equilibrium between stable isotopes and other elements. Conversely, since the Fukushima accident, ^{137}Cs concentrations in konara oak and cedar trees have not reached an equilibrium, and continue to increase (Mahara et al. 2014). This may provide insight into ^{137}Cs dynamics in konara oak trees, but does not present the overall picture.

12.4 Conclusions

Harvesting logs for mushroom cultivation has been an important part of the forestry industry in Fukushima. However, if it is impossible to produce logs with a radioactive Cs level below 50 Bq/kg by using standard production methods, the industry is in a perilous position and must consider significant restructuring. After 100 years, radioactivity will naturally decay to 10 % of current levels; however, the forestry industry cannot wait that long and must continue operations in radioactively contaminated forests despite the associated challenges.

Mushroom log production requires a 20-year cycle, but softwood cedar and cypress used for timber requires at least a 40-year cycle. In Japan, there has been commercial mushroom cultivation by the forestry industry for about 60 years; during previous centuries, hardwood coppices were mostly sources of firewood and charcoal. Remnants of charcoal kilns remain in various forests throughout Japan. Previous forestry practices were long lasting, but during the energy revolution in the 1960s, change happened quickly, and mushroom cultivation is now a major aspect of the forestry industry. In forests with radiation contamination, we must consider implementing historical forestry practices, as forests enrich people's lives in many ways and we should investigate all available options. However, we must also look to the future to make decisions regarding our livelihoods. Fukushima Chuo Forest Cooperative's future management policies are continually developing and depend on further research to understand the problems associated with radioactive contamination of forests.

References

Arkhipov NP, Kuchma ND, Askbrant S et al (1994) Acute and long-term effects of irradiation on pine (*Pinus silvestris*) stands post-Chernobyl. Sci Total Environ 157:383–386. doi:10.1016/0048-9697(94)90601-7

Forestry and Forest Products Research Institute (2012) Results of monitoring radioactive material in stream waters in rainy season. Accessed 15 Nov 2014 (in Japanese)

Forestry Agency (2011) Nation-wide liaison conference of supply and demand of mushroom logs. http://www.rinya.maff.go.jp/j/tokuyou/shiitake/jyukyuukaigi.html. Accessed 20 Nov 2014 (in Japanese)

Forestry Agency (2012a) Trial calculations of exposed dose for a man living in a room surrounded by wood. http://www.rinya.maff.go.jp/j/press/mokusan/pdf/120809_1-02.pdf. Accessed 10 June 2014 (in Japanese)

Forestry Agency (2012b) Revision of index for mushroom logs and culture medium for mushroom beds. http://www.rinya.maff.go.jp/j/press/tokuyou/120328_2.html. Accessed 15 Nov 2014 (in Japanese)

Forestry Agency (2013) Development of a guideline for cultivation management of log-cultured mushrooms to reduce radioactive materials. http://www.rinya.maff.go.jp/j/press/tokuyou/131016.html. Accessed 20 Nov 2014 (in Japanese)

Forestry Agency (2014a) Project of technological development and verification for prevention of radioactive material diffusion from forests. http://www.rinya.maff.go.jp/j/press/kaihatu/140822_1.html. Accessed 20 Nov 2014 (in Japanese)

Forestry Agency (2014b) Survey findings of distribution of radiological materials within forests in fiscal year 2013. http://www.rinya.maff.go.jp/j/press/ken_sidou/pdf/140401-01.pdf. Accessed 10 May 2014 (in Japanese)

Forestry Agency (2014c) Change of major domestic forest products and mushrooms. http://www.e-stat.go.jp/. Accessed 15 Dec 2014 (in Japanese)

Forestry Agency (2014d) Stats of forestry: forest products, lumber industry. http://www.rinya.maff.go.jp/j/kouhou/toukei/. Accessed 15 Dec 2014 (in Japanese)

Forestry Agency (2014e) About withholding conditions of mushrooms and mountain vegetables. http://www.rinya.maff.go.jp/j/tokuyou/kinoko/syukkaseigen.html. Accessed 15 Dec 2014 (in Japanese)

Forestry Agency (2014f) Annual report on forest and forestry in Japan, fiscal year 2013. Forestry Agency, Ministry of Agriculture, Forestry and Fisheries, Japan

Fukushima Lumber Cooperative Unions (2012) Voluntary standard for radioactive safety. http://www.fmokuren.jp/publics/index/24/. Accessed 25 Sept 2014 (in Japanese)

Hasegawa M, Ito MT, Kaneko S et al (2013) Radiocesium concentrations in epigenic earthworms at various distances from the Fukushima nuclear power plant 6 months after the 2011 accident. J Environ Radioact 126:8–13. doi:10.1016/j.jenvrad.2013.06.006

Hashimoto S, Matsuura T, Nanko K et al (2013) Predicted spatio-temporal dynamics of radiocesium deposited onto forests following the Fukushima nuclear accident. Sci Rep 3:2564. doi:10.1038/srep02564

Hayajiri M (2013) Efforts of a forestry cooperative after radiation disaster: its actual circumstances and challenges. J Coop Study Niji 643:71–79 (in Japanese)

Hayashi S (2013) How radiocesium deposits and moves within a forested area. Natl Inst Environ Stud News 32:7–9 (in Japanese)

Hiyama A, Nohara C, Kinjo S et al (2012) The biological impacts of the Fukushima nuclear accident on the pale grass blue butterfly. Sci Rep 2:570. doi:10.1038/srep00570

International Atomic Energy Agency (2002) Modeling the migration and accumulation of radionuclides in forest ecosystems. IAEA, Vienna, pp 1–127

International Atomic Energy Agency (2006) Environmental consequences of the Chernobyl accident and their remediation: twenty years of experience, Report of the Chernobyl forum expert group "Environment". IAEA, Vienna, p 166

Ishida K (2013) Contamination of wild animals: effects on wildlife in high radioactivity areas of the agricultural and forest landscape. In: Nakanishi TM, Tanoi K (eds) Agricultural implications of the Fukushima nuclear accident. Springer, Tokyo, pp 119–129

Kalač P (2001) A review of edible mushroom radioactivity. Food Chem 75:29–35. doi:10.1016/s0308-8146(01)00171-6

Mahara Y, Ohta T, Ogawa H, Kumata A (2014) Atmospheric direct uptake and long-term fate of radiocesium in trees after the Fukushima nuclear accident. Sci Rep 4:7121. doi:10.1038/srep07121

Ministry of Health, Labor and Welfare (2011) Handling of food contaminated by radioactivity. http://www.mhlw.go.jp/english/topics/foodsafety/dl/food-110317.pdf. Accessed 20 Nov 2014

Ministry of Health, Labor and Welfare (2012) Measures against radioactive contamination of food caused by the accident. http://www.mhlw.go.jp/shinsai_jouhou/shokuhin.html. Accessed 20 Nov 2014

Ministry of Health, Labor and Welfare (2013a) Revision of guideline on prevention of radiation hazards for workers engaged in decontamination works. http://www.mhlw.go.jp/english/topics/2011eq/workers/dr/dr/ri_1226_21.html. Accessed 20 Nov 2014

Ministry of Health, Labor and Welfare (2013b) Survey of dietary intake of radionuclides (September to October 2012). http://www.mhlw.go.jp/english/topics/2011eq/dl/index_food_policies_20131011_2.pdf. Accessed 20 Nov 2014

Murakami M, Ohte N, Suzuki T et al (2014) Biological proliferation of cesium-137 through the detrital food chain in a forest ecosystem in Japan. Sci Rep 4:3599. doi:10.1038/srep03599

Neda H (2013) Reducing radiocesium uptake by cultured mushrooms. Tokusan Jouhou 35:8–12 (in Japanese)

Nihei N, Tanoi K, Nakanishi TM (2015) Inspections of radiocesium concentration levels in rice from Fukushima prefecture after the Fukushima Dai-ichi nuclear power plant accident. Sci Rep 5:8653. doi:10.1038/srep08653

Nuclear Regulation Authority (2014) Environmental radioactivity and radiation in Japan. http://search.kankyohoshano.go.jp/servlet/search.top. Accessed 15 Dec 2014 (in Japanese)

Nuclear Regulation Authority (2013) Fundamental perspectives for countermeasures for safety and relief towards the returns of residents (draft). https://www.nsr.go.jp/committee/kisei/h25fy/data/0032_01_1.pdf. Accessed 20 Nov 2014 (in Japanese)

Shcheglov AI, Tsvetnova OB, Klyashtorin AL (2001) Biogeochemical migration of technogenic radionuclides in forest ecosystems. Nauka, Moscow, p 235

Takahashi M (2013) Contamination of radioactive cesium in the forest ecosystem in Fukushima and the necessity of long-term monitoring. Trends Sci 18:68–71 (in Japanese)

Valentin J (2007) The 2007 recommendations of the international commission on radiological protection. Ann ICRP 37:1–332

Yamada T (2013) Mushrooms: radioactive contamination of widespread mushrooms in Japan. In: Nakanishi TM, Tanoi K (eds) Agricultural implications of the Fukushima nuclear accident. Springer, Tokyo, pp 163–176

Yoshida S, Muramatsu Y, Dvornik AM et al (2004) Equilibrium of radiocesium with stable cesium within the biological cycle of contaminated forest ecosystems. J Environ Radioact 75:301–313. doi:10.1016/j.jenvrad.2003.12.008

Chapter 13
Radiocesium in Timber of Japanese Cedar and Japanese Red Pine, in the Forests of Minamisoma, Fukushima

Masaya Masumori, Norio Nogawa, Shin Sugiura, and Takeshi Tange

Abstract The distribution of radiocesium within trees in the forests of Mimamisoma, Fukushima, Japan, was studied between 2012 and 2013 after the Fukushima Nuclear Power Plant accident. Most of the radiocesium was contained in the foliage and bark of the examined trees of Japanese cedar (*Cryptomeria japonica*) and Japanese red pine (*Pinus densiflora*), although considerable concentrations were detected in the xylem of *C. japonica*. At higher positions in the trunk, there was more radiocesium in heartwood than in sapwood. Radiocesium in the xylem of a tree with its root system removed before the nuclear accident suggests that most of the radiocesium was not transferred through the root system but was likely translocated via the foliage.

Keywords Softwood species • Xylem • Radiocesium contamination • Nuclear power plant accident

13.1 Introduction

The Fukushima Daiichi Nuclear Power Plant accident in March 2011 caused massive emissions of radioactive substances into the atmosphere and subsequently over a wide area of forests. Although many reports have examined nuclear materials within trees after the accident (Kuroda et al. 2013; Akama et al. 2013; Ohashi et al. 2014), the number of samples has been limited, and an accurate understanding requires an increased sample size. Since 2012, we have been measuring the radiocesium concentrations in trees in Minamisoma City, north of the nuclear power plant. The measurements have been conducted in cooperation with the

M. Masumori (✉) • S. Sugiura • T. Tange
Laboratory of Silviculture, Graduate School of Agricultural and Life Sciences, The University of Tokyo, 1-1-1, Yayoi, Bunkyo-ku, 113-8657 Tokyo, Japan
e-mail: masumori@fr.a.u-tokyo.ac.jp

N. Nogawa
Fukushima Future Center for Regional Revitalization, Fukushima University, Fukushima 960-1296, Japan

T.M. Nakanishi, K. Tanoi (eds.), *Agricultural Implications of the Fukushima Nuclear Accident*, DOI 10.1007/978-4-431-55828-6_13

Minamisoma City Office and the Soso District Agriculture and Forestry Office, Fukushima Prefecture. The city covers an area of 399 km^2, of which 218 km^2 is covered by forest. Approximately half of the forest is artificially planted for timber production. This study concentrated on two timber tree species, Japanese cedar (*Cryptomeria japonica*), and Japanese red pine (*Pinus densiflora*), which make up 3/4 of the standing timber volume of the forest.

13.2 Study Sites and Measurement of Radiocesium

We investigated five forests stands owned and managed by Minamisoma City, locating 20–35 km NNW of the Fukushima Daiichi Nuclear Power Plant. All stands contained 50–60-year-old plantation forests.

According to the airborne monitoring map by Ministry of Education, Culture, Sports, Science and Technology, the minimum level of cesium 137 (^{137}Cs) at all sites was 300 kBq/m^2 and the maximum was estimated to be up to 3000 kBq/m^2 on April 29, 2011. As an indicator of radiocesium deposition at each site at the time of sampling, we measured the air dose rate 1 m above the ground and close to the trees using a NaI scintillation survey meter.

The radionuclides were quantified in each sample using a germanium semiconductor detector. Peaks corresponding to ^{134}Cs and ^{137}Cs were detected for each sample. As ^{134}Cs decayed naturally from 2012 to 2013, ^{137}Cs levels were used in the present chapter to examine the radioactivity over the 2 years.

13.3 Distribution of Radiocesium in Standing Trees

Three trees, two *C. japonica* and one *P. densiflora*, were cut down in December 2012 (21 months after the accident) and in December 2013 (33 months after the accident), and the distribution of radiocesium on the inside and outside of the trees was investigated (Table 13.1). Among the six trees, five trees (#4 to #8) were felled from the same stand, and a *C. japonica* tree (#3) was felled from a stand with a comparatively higher air dose rate. Both stands were on southern facing slopes and were thinned in 2009.

To prevent soil particles contaminating the trunk after felling, each trunk was covered with a plastic sheet to a height of 1.5 m and the tree was cut close to the ground (Fig. 13.1A). After cutting, branches and foliage that did not touch the soil were sampled. The sample logs for analysis were taken at seven positions of each trunk in 2012 and at 2–3 positions in 2013 (Fig. 13.1B). The whole tree was weighed, including the parts not sampled.

The sample logs were transported to a sawmill and cut into 5 cm disks with a bandsaw. The logs harvested in 2012 were subsequently separated into bark, heartwood, and sapwood. The logs harvested in 2013 were stripped of the bark at

Table 13.1 Radiocesium content in trunks of standing trees (Masumori et al. 2015)

#	Dose rate at the site of the day	Tree height Height to crown Diameter at 1.3 m	Disc height	^{137}Cs content Bq/g Bark	Sapwood	Transition wood	Heartwood
#3 *Cryptomeria*	3.6 μSv/h	21.7 m 14.0 m 28 cm	19 m	11.77	0.62	←	–
			16 m	7.34	0.43	←	1.38
			13 m	16.97	0.52	←	1.00
			10 m	10.90	0.51	←	0.92
			7 m	13.86	0.62	←	0.77
			4 m	13.47	0.49	←	0.55
			1.3 m	15.28	0.72	←	0.65
			−0.2 m	1.08	0.30	←	←
			−0.5 m (lateral root)	0.96	0.23	←	←
#4 *Cryptomeria*	1.7 μSv/h	23.2 m 15.7 m 34 cm	19 m	8.58	0.35	←	0.69
			16 m	4.71	0.35	←	0.46
			13 m	5.11	0.41	←	0.39
			10 m	4.76	0.38	←	0.37
			7 m	3.25	0.31	←	0.26
			4 m	3.16	0.29	←	0.24
			1.3 m	2.43	0.30	←	0.32
#5 *Cryptomeria*	1.3 μSv/h	24.0 m 13.3 m 39 cm	19 m	2.88	0.18	0.41	1.04
			10 m	2.69	0.27	0.50	0.61
			1.3 m	1.96	0.36	0.46	0.49
#6 *Cryptomeria*	1.3 μSv/h	15.8 m 6.4 m 19 cm	10 m	5.16	0.28	0.43	0.90
			1.3 m	2.27	0.27	0.61	0.98
#7 *Pinus*	1.8 μSv/h	22.2 m 14.6 m 28 cm	19 m	1.10	0.20	←	–
			16 m	1.30	0.20	←	–
			13 m	1.20	0.20	←	0.10
			10 m	1.40	0.20	←	0.10
			7 m	1.50	0.20	←	0.10
			4 m	1.30	0.10	←	0.10
			1.3 m	3.40	0.20	←	0.10
			−0.4 m (lateral root)	1.04	0.19	←	←
			−1 m (lateral root)	2.39	0.38	←	←
#8 *Pinus*	1.3 μSv/h	20.3 m 15.6 m 21 cm	16 m	1.00	0.11	←	0.11
			10 m	0.93	0.61	←	0.10
			1.3 m	1.40	0.12	←	0.06

Heartwood and transition wood were measured separately for *Cryptomeria japonica* #5 and #6. Heartwood had not formed near the crown apex in *C. japonica* #3 and *Pinus densiflora* #7. Heartwood and transition wood of the root samples were not separated for analysis

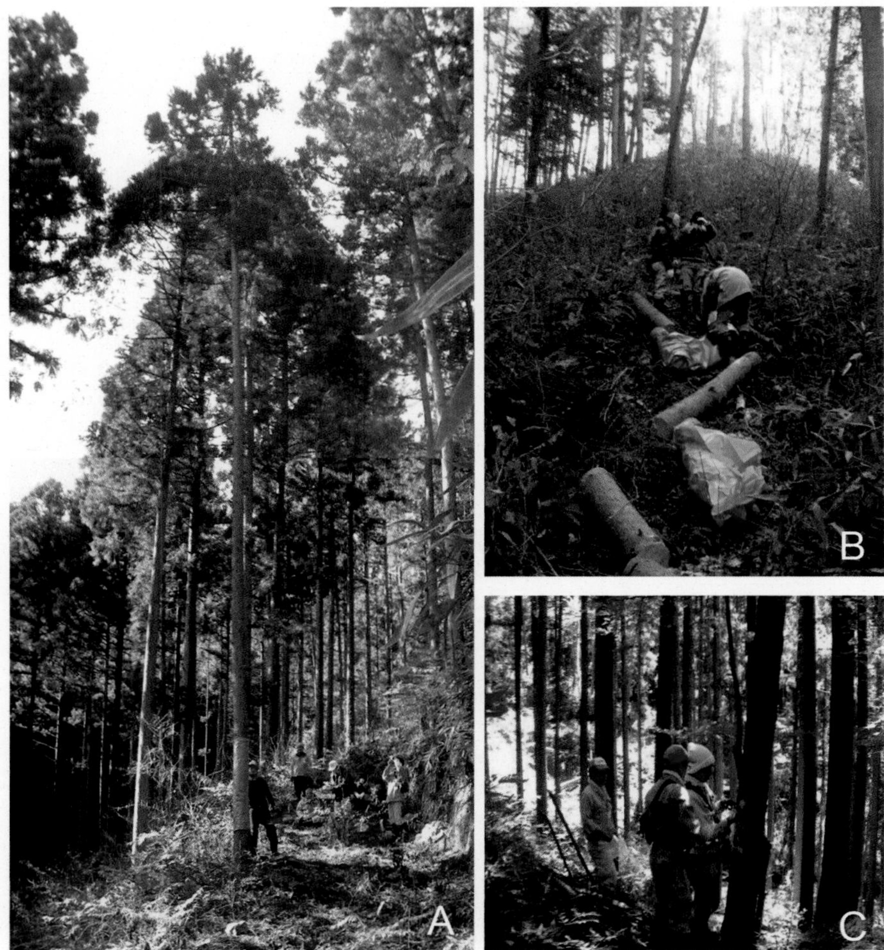

Fig. 13.1 Tree sampling. (**A**) Preparing to fell a *Cryptomeria japonica* tree. (**B**) Wrapping sample logs of *Pinus densiflora* in vinyl sheets to avoid contact with contaminated soil. (**C**) Drilling out xylem chips from a standing *C. japonica* tree

the felling site, samples of the bark were taken, and the remaining bark fragments were brushed away. Subsequently, disks composed of xylem were cut and divided into sapwood and heartwood. In addition, for a *C. japonica* sample in 2013, the pale-colored rings between sapwood and heartwood were further sampled separately as transition wood. The woods were cleaved into 5-cm long and 0.2–1-cm thick fragments, dried, and analyzed to determine their ^{137}Cs levels.

The stumps of *C. japonica* #3 and *P. densiflora* #7 were pulled out of the ground with a grapple machine. Lateral roots of 4–7 cm diameter were sampled, and the xylem with the bark removed was analyzed for ^{137}Cs.

Table 13.2 Radiocesium in foliage

^{137}Cs content Bq/g									
#	Grown in 2013		Grown in 2012		Grown 2011		Grown before 2010		Older shoot
	Needle	Stem	Needle	Stem	Needle	Stem	Needle	Stem	
#3 *Cryptomeria*			14.16		38.87		74.50		39.57
#4 *Cryptomeria*			5.42		11.67		13.73		10.47
#5 *Cryptomeria*	2.04		1.51		2.45		3.66		4.90
#6 *Cryptomeria*									
Branch 8	0.93		3.80		4.27		9.90		5.89
Branch 7	0.67		0.71		5.33		7.08		4.02
Branch 6	0.54		0.90		2.22		7.70		2.54
Branch 5	0.53		0.63		1.75		4.14		4.63
Branch 4	1.02		1.09		3.90		1.02		3.62
Branch 3	0.89		1.24		4.44		6.97		4.39
Branch 2	1.81		2.55		4.27		9.12		4.64
Branch 1	1.60		2.26		7.53		9.00		3.72
#7 *Pinus*			1.52	2.62	1.40	8.55	→		17.16
#8 *Pinus*	0.89	0.62	0.38	0.62	–	1.14	–	2.71	2.16

Cryptomeria japonica: leaves and stems covered with leaves were not separated for analysis. *Pinus densiflora*: no needles were attached on older shoots more than 1 year old

There was a large variation in radiocesium levels among the trees and within each tree depending on the sampled positions. The highest concentrations occurred in the foliage followed by the bark of the trunk (Tables 13.1 and 13.2). Radiocesium was detected in the xylem of all logs (Table 13.1).

The total volume of radiocesium in each organ was calculated by multiplying the radiocesium concentration by the dry weight. In 2012, less than 20 % of the radiocesium from the aboveground part was in the trunk; most was in the foliage. In 2013, up to 40 % of radiocesium was in the trunk. The change in the distribution of radiocesium was probably due to gradual defoliation of foliage that was on the trees at the time of the accident.

To increase the sample size, wood samples were taken without felling from trunks with a band drill from 11 *C. japonica*, three *P. densiflora*, and one Japanese cypress (*Chamaecyparis obtusa*) at three stands.

A 7 cm × 7 cm sample of bark was peeled from each sample tree at a breast height using a knife and a handsaw. The exposed xylem was brushed and then drilled in the radial direction with a hand drill of an 18 mm drill bit and a clickball. The shavings from the drilling operation were collected in a plastic bag (Fig. 13.1C). The sapwood from *C. japonica* and *C. obtusa* was collected until the shavings became darker, and subsequent shavings were heartwood samples. The heartwood and sapwood of *P. densiflora* was harder to distinguish on the basis of color; therefore, both were collected together without separation. Three to five holes were drilled to provide sufficient samples for analysis.

The radiocesium concentration was high in the bark of all trees and it was also detected in the xylem. Although the radiocesium levels in the trees tended to be higher in the stand with the higher air dose rate, there was also a large variation between the trees in each stand (Fig. 13.2). It may be possible to estimate the amount of radiocesium in the xylem from the air dose rate and to estimate the cesium deposition, but the large variation between trees should be noted.

The distribution of radiocesium within 5-mm thick disks cut from the same logs used for quantitative measurement of radiocesium was visualized using an imaging plate (BAS-IP MS, GE Healthcare Japan). Disks and image plates were stacked in the cassette and kept in a box made of ≥5-cm thick lead bricks to shield against natural radiation. Images were obtained using a fluorescent image analyzer (FLA-9000, Fujifilm) after the exposure for 1–5 months. Disks were also taken from *C. japonica* and *P. densiflora* trees felled at the Ecohydrology Research Institute, University of Tokyo, 425 km southeast of the Fukushima Daiichi nuclear power plant, and were exposed to the image plates under the same conditions as described above. However, no images could be detected from these log disks. This indicates that the images obtained from the Minamisoma samples represented the distribution of radioactive substances emitted from the nuclear power plant accident.

The images from each log disk are shown in Fig. 13.3. Similar to the results from Table 13.1, *C. japonica* have higher radiocesium concentrations in the heartwood than in the sapwood, especially for disks that were taken from higher positions nearer the crown. Kuroda et al. (2013) reported higher concentrations in *C. japonica* sapwood than heartwood, but they only analyzed wood lower than 3 m; therefore, their results do not necessarily conflict with ours. The imaging plates for the trunk xylem of *C. japonica* #5 show a dark color indicating the strong presence of radiocesium, which appeared to be highest at the outer edges of heartwood where heartwood formation was taking place (Fig. 13.3). For *C. japonica* #5 and #6, we analyzed transition wood with a low moisture content at the boundary between the heartwood and sapwood. The dry weight concentration of radiocesium was higher in the heartwood, followed by the transition wood, and then the sapwood (Table 13.1). In *C. japonica*, rubidium is actively transported from the sapwood to the outer heartwood via xylem ray (Okada et al. 2012), indicating that this mechanism could be used to transport another alkaline metals such as cesium.

The concentration of radiocesium was less in the xylem of *P. densiflora* than *C. japonica* and less in the heartwood than the sapwood of both the felled trees and the cored samples (Fig. 13.2). Samples of *C. obtusa*, another timber species, were taken from only one tree, but the radiocesium distribution in the xylem was similar to *C. japonica*. If the radiocesium was initially absorbed into the xylem via the foliage, then there should be a relationship between the foliage volume and radiocesium amount in the xylem. For the five trees felled from the same stand (#4–#8), the total radiocesium amount in the xylem of each trunk was compared with the dry weight of the foliage (as an index of total foliage volume) to test whether there was a correlation. The radiocesium amount in the trunk xylem was proportional to the 2/3 power of foliage dry weight, which indicates a surface area

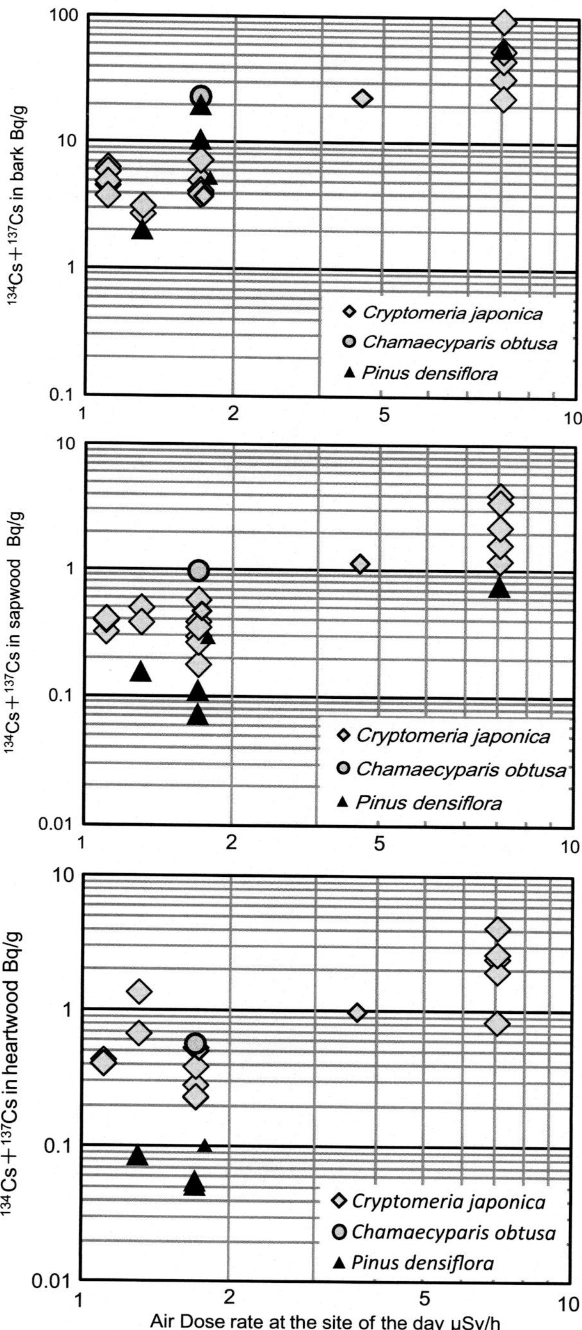

Fig. 13.2 The radiocesium content of trunks taken from a number of forest stands with different air dose rates (Masumori et al. 2015). Data from Table 13.2 is included. Small symbols: trees felled in 2012

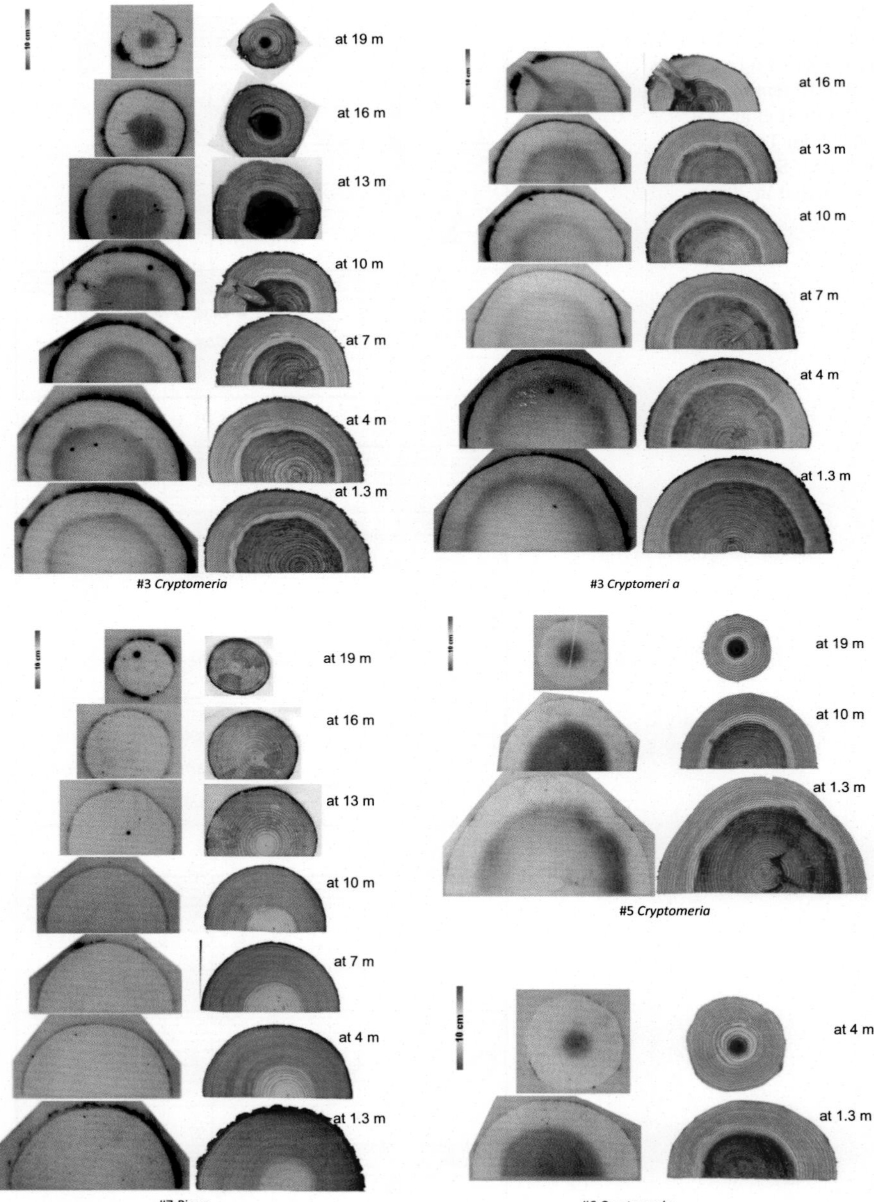

Fig. 13.3 Trunk cross-sections showing distribution of radioactivity. Five sample trees cut into log disks at various heights (*right*) were exposed to imaging plates (Masumori et al. 2015). *Darker colors* indicate higher radioactivity. The log disks included the bark for #3, #4, and #7. The *black spots* on the xylem image are due to some scattering of bark fragments. The bark from *C. japonica* #5 and #6 was peeled off before cutting the log disks; therefore, there is no interference from bark fragments. The bark from *P. densiflora* #8 was stripped off in the same way, but no radiation was seen (data not shown)

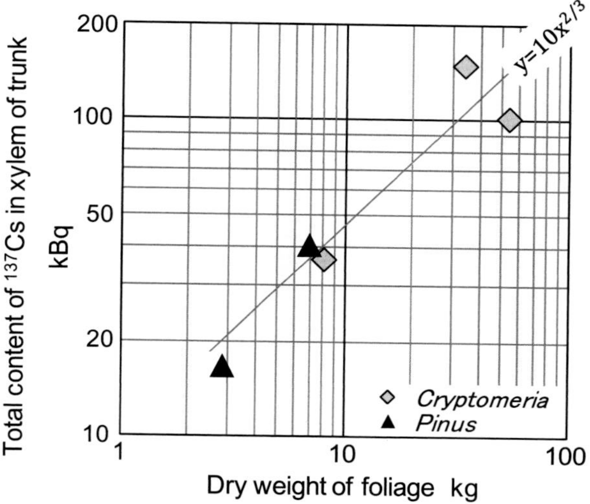

Fig. 13.4 Total radiocesium content in the xylem of trees and the dry weight of the foliage from a single stand (Masumori et al. 2015)

(Fig. 13.4). The interspecific differences in radiocesium content between the xylem of *C. japonica* and *P. densiflora* were not related to the anatomical or physiological characters of each species but were more likely due to differences in the areal quantity of the foliage surface where radiocesium deposited.

13.4 Distribution of Radiocesium in the Crown

Radiocesium was analyzed in the leaves of different ages. From the felled *P. densiflora* trees, we separately sampled the needles from current-year shoots and needles from shoots elongated in the previous year. On the sampling day in December, the pine had no needles on the older shoots. Young *C. japonica* shoots are densely covered with needle-like leaves; thus, the leaves and stem were not separated. The remaining shoots comprising leaves and stem were separated according to the year of growth, up to 3 years old. Shoots more than 3 years old were counted as branches, even if they still had needles attached.

Branches that developed before the accident in March 2011 showed the highest radiocesium concentrations, although radiocesium was detected in foliage that had developed after the accident (Table 13.2).

Eight branches were analyzed separately from *C. japonica* #5. A large variation in radiocesium concentration occurred between the branches and the foliage (Table 13.2). In agreement with Akama et al. (2013), we found that for each branch, the younger leaves tended to have the lowest radiocesium concentrations, but the concentrations in the older leaves did not correlate with those in the younger leaves (Table 13.2). Therefore, the movement of radiocesium to newly elongating shoots

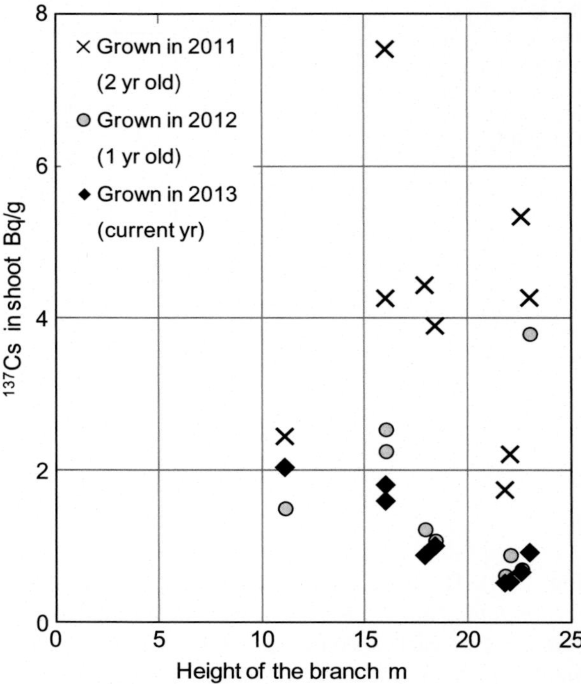

Fig. 13.5 Radiocesium content in *C. japonica* leaves developed after the nuclear accident. Data from Table 13.3 is included. The value at height of 11.1 m is for Tree #6. All other values are for Tree #5

of the same branch was not necessarily driven by a concentration gradient. The youngest leaves on branches at a lower position on the tree had higher concentrations of radiocesium than those on higher branches. The youngest leaves of *C. japonica* #6 had lower positioned foliage and higher radiocesium concentration than those of *C. japonica* #5 (Table 13.2, Fig. 13.5). Although most radiocesium in the older foliage was immobilized, some radiocesium is likely to be mobile and translocated not only within the tree but also outside the tree and downwards through the forest canopy such as with rain. We suggest that the position in the canopy should be taken into account when considering radiocesium migration into developing organs.

13.5 Radiocesium in Xylem from Fallen Trees

In August 2012, samples were taken from the trunks of two *C. japonica* trees that had already fallen and their radiocesium concentrations were measured (Table 13.3). At the forest site where the two trees had grown, thinning was occurring at the time of the earthquake on March 11, 2011. Tree #1 had been felled, but work was interrupted because of the earthquake before limbing procedure and the tree was left on the forest floor with the foliage still attached. Tree #2 had been growing in the vicinity of tree #1 but had not been thinned, although it was uprooted during the typhoon in June 2012 and had been lying on the forest floor for a month and a half at the time of our research. Consequently, tree #1 would have been separated from its roots at the time of the nuclear power plant accident, but tree #2 would have been intact and growing.

Two disks were sampled from the logs of each tree at 1.3 m and 13 m from the ground. At these positions, the trunk had not touched the soil, even after the tree had fallen. From each disk, the bark was sampled and the xylem was separated into six fractions from the sapwood to the pith according to the tree rings. For tree #1, separate cesium analyses were conducted for the semicircular half of lying trunk facing the sky and the semicircular half of the lying trunk facing the ground.

Radiocesium was present in the xylem of the trunk of *C. japonica* #1, which had been separated from its roots at the time of the nuclear power plant accident. *C. japonica* #2 had similar radiocesium concentrations in the xylem, although it had been an intact tree up to 15 months after the accident (Table 13.3). This

Table 13.3 Radiocesium content in trunk of fallen trees

#	Tree height Diameter at 1.3 m	Discat	Radial Position	^{137}Cs content Bq/g Skyward semi circle beneath semi circle	
#1 *Crvotomeria*	25.0 m	13 m	Bark	12.14	8.80
	26 cm		7 ~ 9 cm	0.58	0.63
			4 ~ 7 cm	0.28	0.48
			Pith ~ 3 cm	0.18	0.27
		1.3 m	Bark	7.88	3.71
			11 ~ 12 cm	0.24	0.18
			9 ~ 11 cm	0.16	0.12
			7 ~ 9 cm	0.08	0.06
			5 ~ 7 cm	0.06	0.04
			3 ~ 5 cm	0.04	0.02
			Pith ~ 3 cm	0.09	0.01
#2 Crvntomeria	23.0 m	13 m	Bark	44.23	
	20 cm		8 ~ 9 cm	0.38	
			7 ~ 8 cm	0.26	
			6 ~ 7 cm	0.2	
			5 ~ 6 cm	0.22	
			3 ~ 5 cm	0.19	
			Pith ~ 3 cm	0.15	
		1.3 m	Bark	5.64	
			5 ~ 6 cm	0.2	
			4 ~ 5 cm	0.17	
			3 ~ 4 cm	0.31	
			Pith ~ 3 cm	0.54	

For *Cryptomeria japonica* #2, both semicircle samples were pooled for measurement suggests that most of the radiocesium in the xylem of *C. japonica* is not absorbed through the roots. The concentration of radiocesium in the roots of *C. japonica* #3 was lower than that in the trunk (Table 13.1), suggesting that a comparatively small amount of radiocesium had migrated from the roots to the trunk at the time of sampling.

If the radiocesium emitted by the accident had precipitated with the rain, the amount of radiocesium deposited on the sky-facing and ground-facing of the fallen *C. japonica* #1 should differ. Higher concentrations of radiocesium were detected in the sky-facing bark (Table 13.3). If radiocesium diffused inward from the bark to the xylem, there should be a higher concentration in the sky-facing xylem, but in contrast to the bark, there were no clear differences between sky-facing and ground-facing xylem (Table 13.3). Although *C. japonica* #1 was felled, the foliage remained in place; thus, physiological activity would have continued at the tissue level. Therefore, the radiocesium detected in the xylem of the trunk may have migrated basipetally through the vascular bundle from deposits on the foliage.

13.6 Greenhouse Experiments

Greenhouse cultivation experiments were conducted to determine the characteristics of radiocesium absorption from the roots in both *C. japonica* and *P. densiflora*. These experiments also allowed us to examine the effects of secondary deposits from the canopy, which contained a high quantity of radioactive deposition, and migration from resuspended soil particles.

Two-year-old seedlings grown in a radiocesium-free environment were transplanted in pots with soil from Minamisoma forests containing 20 Bq/g of ^{137}Cs. After 5 months in the greenhouse, shoots from three *C. japonica* seedlings and seven *P. densiflora* seedlings were analyzed for radiocesium. Because a 5-month period was insufficient for the *C. japonica* plants to recover an appropriate contact between root system and potting media, not much shoot growth was seen during this period. ^{137}Cs migration into the *C. japonica* shoots was 5–10 Bq/kg, and migration into the *P. densiflora* shoots was 2–38 Bq/kg. Imaging plates show that the radioactivity was uniformly distributed among all the organs within the shoots, except it was higher in newly grown *P. densiflora* shoots (Fig. 13.6).

In standing trees, the incorporated radiocesium in the xylem was greater in *C. japonica* than in *P. densiflora*, while migration through the roots did not differ between the species. In these forests, the levels of radiocesium in the xylem can gradually increase because of absorption via the root from the forest floor where much of the radiocesium is deposited and is accumulating.

Fig. 13.6 Radiocesium in seedlings absorbed through the roots. *Cryptomeria japonica* (*top*) and *Pinus densiflora* (*bottom*) (Masumori et al. 2014). The seedlings were planted in a greenhouse in potting medium rich in radiocesium (*right panels*). The shoots were exposed to imaging plates 5 months later to visualize the radiocesium distribution (*left panels*)

References

Akama A, Kiyono Y, Kanazashi T, Shichi K (2013) Survey of radioactive contamination of sugi (*C. japonica* japonica D. Don) shoots and male flowers in Fukushima prefecture. Jpn J For Environ 55:105–111

Kuroda K, Kagawa A, Tonosaki M (2013) Radiocesium concentrations in the bark, sapwood and heartwood of three tree species collected at Fukushima forests half a year after the Fukushima Dai-ichi nuclear accident. J Environ Radioact 122:37–42

Masumori M, Nogawa N, Tange T (2014) Distribution of radiocesium released from the NPP accident within trees – cases of forests in Minamisoma, Fukushima. Tree For Health 18:70–71 (in Japanese)

Masumori M, Nogawa N, Sugiura S, Tange T (2015) Radiocesium in stem, branch and leaf of *Cryptomeria japonica* and *Pinus densiflora* trees: cases of forests in Minamisoma in 2012 and 2013. J Jpn For Soc 97:51–56 (in Japanese with English summary)

Ohashi S, Okada N, Tanaka A, Nakai W, Takano S (2014) Radial and vertical distributions of radiocesium in tree stems of *P. densiflora* densiflora and *Quercus serrata* 1.5 y after the Fukushima nuclear disaster. J Environ Radioact 134:54–60

Okada N, Hirakawa Y, Katayama Y (2012) Radial movement of sapwood-injected rubidium into heartwood of Japanese cedar (*C. japonica* japonica) in the growing period. J Wood Sci 58:1–8

Chapter 14
Ecosystem Monitoring of Radiocesium Redistribution Dynamics in a Forested Catchment in Fukushima After the Nuclear Power Plant Accident in March 2011

Nobuhito Ohte, Masashi Murakami, Izuki Endo, Mizue Ohashi, Kohei Iseda, Takahiro Suzuki, Tomoki Oda, Norifumi Hotta, Keitaro Tanoi, Natsuko I. Kobayashi, and Nobuyoshi Ishii

Abstract The accident at the Fukushima Daiichi Nuclear Power Plant in March 2011 emitted 1.2×10^{16} Bq of cesium-137 (^{137}Cs) into the surrounding environment. Radioactive substances, including ^{137}Cs, were deposited onto forested areas in the northeastern region of Japan. ^{137}Cs is easily adsorbed onto clay minerals in the soil; thus, a major portion of ^{137}Cs can be transported as eroding soil and particulate organic matter in water discharge. Dissolved ^{137}Cs can be taken up by microbes, algae, and plants in soil and aquatic systems. Eventually, ^{137}Cs is introduced into insects, worms, fishes, and birds through the food web. To clarify the mechanisms of dispersion and export of ^{137}Cs, within and from a forested ecosystem, we conducted intensive monitoring on the ^{137}Cs movement and storage in a forested headwater catchment in an area approximately 50 km from the Nuclear Power Plant. Two major pathways of ^{137}Cs transport are as follows: (1) by moving water via dissolved and particulate or colloidal forms and (2) by dispersion through the food web in the forest-stream ecological continuum. The ^{137}Cs concentrations of stream waters were monitored. Various aquatic and terrestrial organisms were

N. Ohte (✉)
Graduate School of Informatics, Kyoto University, Kyoto, Japan
e-mail: nobu@bre.soc.i.kyoto-u.ac.jp

M. Murakami • T. Suzuki
Graduate School of Science, Chiba University, Chiba, Japan

I. Endo • K. Iseda • T. Oda • K. Tanoi • N.I. Kobayashi
Graduate School of Agricultural and Life Sciences, The University of Tokyo, Tokyo, Japan

M. Ohashi
School of Human Science and Environment, University of Hyogo, Himeji, Japan

N. Hotta
Graduate School of Life and Environmental Sciences, University of Tsukuba, Tsukuba, Japan

N. Ishii
National Institute of Radiological Sciences, Chiba, Japan

T.M. Nakanishi, K. Tanoi (eds.), *Agricultural Implications of the Fukushima Nuclear Accident*, DOI 10.1007/978-4-431-55828-6_14

periodically sampled to measure their ^{137}Cs concentrations. The results indicate that the major form of exported ^{137}Cs is via suspended matter. Particulate organic matter may be the most important carrier of ^{137}Cs. High water flows generated by a storm event accelerated the transportation of ^{137}Cs from forested catchments. Estimation of ^{137}Cs export from the forested catchments requires precise evaluation of the high water flow during storm events. The results also suggested that because the biggest pool of ^{137}Cs in the forested ecosystem is the accumulated litter and detritus, ^{137}Cs dispersion is quicker through the detritus food chain than through the grazing food chain.

Keywords Cs • Forest ecosystem • Suspended solid • Food web • Bioaccumlation

14.1 Introduction

Following the Fukushima Daiichi Nuclear Power Plant accident in March 2011, approximately 1.5×10^{17} Bq of iodine-131 (^{131}I) and 1.2×10^{16} Bq of cesium-137 (^{137}Cs) were emitted into the surrounding environment (Ministry of Education, Culture, Sports, Science and Technology and Ministry of Agriculture, Forestry and Fisheries 2012). These radioactive substances were deposited on the northeastern region of Honshu Island, Japan. Forests occupy more than 70 % of the total area in these regions and are particularly important for local societies, not only because of their association with the forestry industry but also for their residential environment, which includes source areas of drinking water. The first phase of government surveys revealed that a major portion of the deposited radiocesium was trapped in the canopy and in the litter layer on the forest floor (Ministry of Education, Culture, Sports, Science and Technology; Ministry of Agriculture, Forestry and Fisheries 2012; Hashimoto et al. 2012). Radiocesium has been shown to be easily adsorbed by clay minerals and soil organic matter (Kruyts and Delvaux 2002), which can be transported by eroded soil, particulates, and dissolved organic matter through hydrological channels, streams, and rivers (e.g., Fukuyama et al. 2005; Wakiyama et al. 2010). Within the forested ecosystem, radioactive materials deposited on the tree canopies subsequently move to the forest floor by precipitation (Kinnersley et al. 1997; Kato et al. 2012) and litter fall (Bunzl et al. 1989; Schimmack et al. 1993; Hisadome et al. 2013). It is expected that the movement of radiocesium from the canopy to the forest floor will gradually decrease (Hashimoto et al. 2013), and it is considered that its movement in the Fukushima forests has been active because only 3.5 years have passed since the accident.

Dissolved radiocesium, which is relatively free from soil adsorption, can also be taken up by microbes, algae, and plants in soil and aquatic ecosystems. By propagating through the food web in the forest ecosystem, it was expected that radiocesium would eventually be introduced into soil, insects, worms, fishes, and birds. Many previous reports on the distribution and transfer of radionuclides have focused on bioaccumulation and the transition between trophic levels (Kitchings et al. 1976; Rowan and Rasmussen 1994; Wang et al. 2000).

Therefore, it is necessary to investigate the major pathways of radiocesium transfer in the forest, such as physical transportation via hydrological processes and biological transfer through the food web in terrestrial and aquatic ecosystems. To describe the current status and to examine the hypothesized mechanisms mentioned above, catchment scale biogeochemical observations have been conducted in the northern part of the Fukushima prefecture since August 2012 (Ohte et al. 2012, 2013), and they currently continue (February 2015). Here, we report the latest results and discuss the future perspectives for monitoring radiocesium levels in Fukushima forests.

14.2 Materials and Methods

14.2.1 Study Site

The study area is the Kami-Oguni River catchment, which is located approximately 50 km from the Fukushima Daiichi Nuclear Power Plant. The Ministry of Education, Culture, Sport, Science and Technology (2013) used aircraft survey devices to determine that the air dose rate in this region was 1.0–1.9 μSv h^{-1} and the total deposition rate of ^{137}Cs was 300–600 kBq m^{-2} in 2013. To intensively monitor the radioactive cesium dynamics, we set a small sub-catchment in a forest ecosystem in the headwater part of the Kami-Oguni River catchment, known as the Sanbo Observational Forest Catchment.

The geology of the catchment is dominated by volcanic rocks (andesite and basalt) formed by volcanic eruptions during the Miocene Epoch. The soil is classified as Dystrudepts (Soil Survey Staff 2014), which is characterized by high exchangeable Mg^{2+} and low K^{+} concentrations. The soil texture in the topsoil (A-Horizon) of this area was a mixture of clay < 0.002 mm (25.3 %), silt 0.02–0.002 mm (40.1 %), sand 2–0.02 mm (34.6 %) determined with the sieving method (Kato et al. unpublished data).

Major areas of Sanbo Observational Forest Catchment are covered by typical secondary stands that consist of broadleaf deciduous species dominated by *Quercus serrata* Murray and *Zelkova serrata*. Some parts of these secondary stands were mixed with naturally regenerated Japanese red pine (*Pinus desiflora*). The valley areas of the catchment were used as Japanese cedar (*Cryptmeria japonica*) plantations for timber production. *Pinus desiflora* and *Cryptmeria japonica* are evergreen conifers.

14.2.2 Field Observations and Sampling

To monitor the fluxes of radiocesium transported by water flow into and out of the forest catchment, we measured water fluxes and radiocesium concentrations for each elemental hydrological process, such as rainfall, throughfall, stemflow, and stream water discharge. The water discharge from the catchment was continuously measured using a partial flume with a water level recorder. Stream water was sampled monthly and intensively during storm events.

Three rectangular plots including two deciduous-pine mixed stands (DP1 and DP2, 20 m × 20 m) and one cedar plantation (CP, 10 m × 40 m) were selected to investigate the spatial distributions and movements of radiocesium in hydrological and biogeochemical processes within the plant and soil system. We conducted monthly samplings of litter fall, throughfall, stemflow, and soils from each plot, and quantified all samples. Details of the sampling are described in Endo et al. (2015).

Food web components including terrestrial and aquatic organisms in the continuum of forest and stream ecosystems were sampled to investigate the current status of radiocesium flows and storages affected by biological activities. All samples were collected from the study site and identified to the genus and species level, and then classified into 10 functional groups according to the criteria shown in Fig. 14.1. The sampling points for each sample were selected randomly within the study catchment. The samples were collected in May, July, and September 2012 and in February 2013. Detailed descriptions of the sampling and data analysis are presented in Murakami et al. (2014).

Whole trees were sampled in November 2012 and 2013 to estimate the stocks of radiocesium in the above ground biomass of the dominant trees, *Quercus serrata* and *Cryptomeria japonica*, at the study site. *Quercus serrata* is the most common species in the secondary deciduous forest in this region, and *C. japonica* is the most common species in the plantations for timber production in Japan. Three

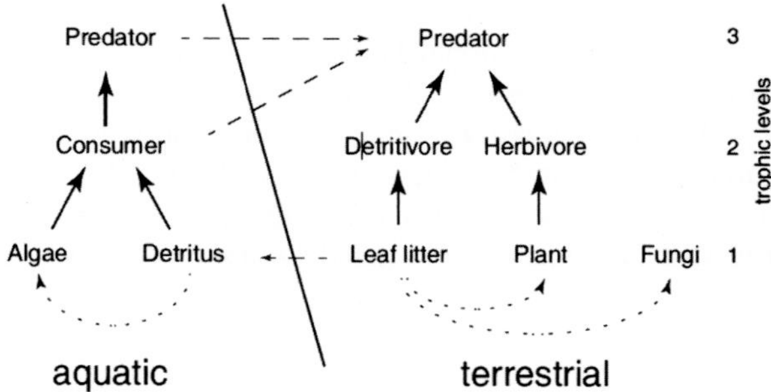

Fig. 14.1 Schematic food web of the present study. Ten functional groups were used as sampling units. *Solid lines* indicate trophic interactions and *dashed lines* indicate nutrient flow. *Broken lines* show spatial movements; e.g., transformation of tadpoles to frogs and dragonflies or supply of leaf litters from forest to stream (Murakami et al. 2014)

individuals of *Q. serrate* and one *C. japonica* were sampled at each sampling event in November of 2012 and 2013. Only a single *C. japonica* was sampled, because these trees were planted at a specific time and thus their tree sizes were homogeneous.

Radiocesium concentrations (^{137}Cs and ^{134}Cs) of leaves, branches, and stems were measured separately. Moreover, stem samples were separated into sapwood and heartwood. Prior to these samplings and measurements, surveys of individual tree sizes were performed for three rectangle plots. All trees with a diameter at breast height (DBH) >5 cm were measured for their DBH and height at the top of the canopy.

14.2.3 Sample Analysis

Germanium semiconductor detectors were used to measure radiocesium concentrations for all samples. Gamma-ray spectrometry was conducted using germanium detectors (Seiko EG&G). The measured values were corrected for the sampling day, and were expressed in Bq kg^{-1} of dry weight for the organic samples and in Bq L^{-1} for the water samples, respectively.

To evaluate the relative trophic levels for the sampled organisms, the stable nitrogen isotope ratio (δ^{15}N) for each sample was measured using SerCon ANCA GSL elemental analyzer interfaced to a SerCon Hydra 20–20 continuous flow isotope ratio mass spectrometer. Analysis and data treatments are explained in Murakami et al. (2014).

14.3 Results and Discussion

14.3.1 ^{137}Cs Concentrations of Tree Parts

The leaf ^{137}Cs concentrations for *C. japonica* were higher than 10,000 Bq kg^{-1} for the mature and current leaves, but they decreased to 3500 Bq kg^{-1} for mature leaves and 2700 Bq kg^{-1} for current leaves (Fig. 14.2). These results indicate that the leaves that were on the canopy when the radiocesium was deposited still contained very high ^{137}Cs levels in November 2012, and even the new leaves in 2012 had similarly high levels of ^{137}Cs concentration. These results suggest that ^{137}Cs was translocated from the old leaves to the new leaves. The decreasing ^{137}Cs concentrations in the mature leaves implies that they were replaced with new leaves with lower ^{137}Cs concentrations, and also that the ^{137}Cs attached to the leaf surfaces of the old leaves was gradually washed out by rainfall.

On the other hand, leaves of *Q. Serrata* had 1000 and 990 Bq kg^{-1} in 2012 and 2013, respectively. Considering that no living leaves were on the canopies of these

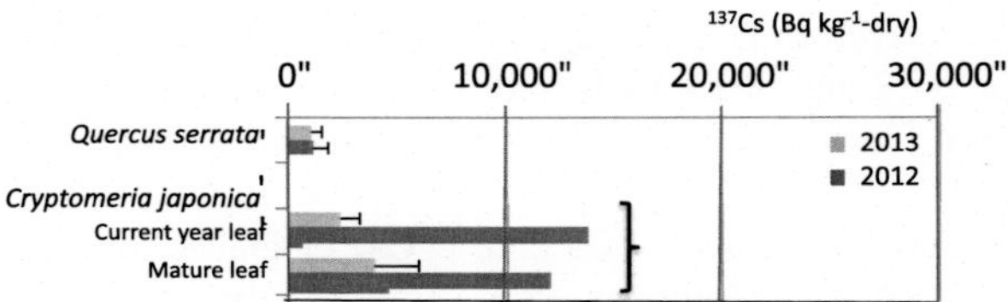

Fig. 14.2 The ^{137}Cs concentration in living leaves of *Quecus serrate* and *Cryptomeria japonica*. Samples were collected from whole tree sampling conducted in early November 2012 and 2013. Because *C. japonica* is an evergreen species, it has multiple aged leaves, while *Q. serrate* is a deciduous species. Data without error bars were derived from the samples without replications

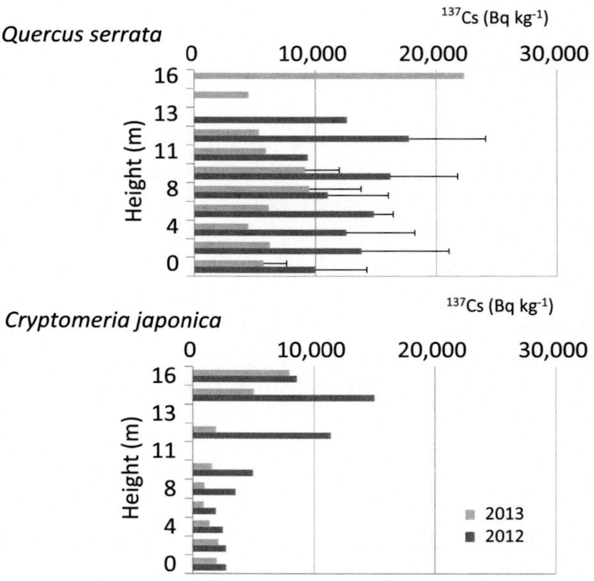

Fig. 14.3 The ^{137}Cs concentration in the trunk bark of *Quecus serrate* and *Cryptomeria japonica*. Samples were collected from whole tree sampling conducted in early November 2012 and 2013. Data without error bars were derived from the samples without replications

deciduous trees when the radiocesium fell in March 2011, the ^{137}Cs in the living leaves was probably transported from the other parts of the tree in 2012 and 2013.

For *C. japonica*, the concentrations of ^{137}Cs in the bark within the evergreen canopy part of the tree (>9000 Bq kg^{-1} in 2012 and 5000–7000 Bq kg^{-1} in 2013) were higher than in the bark beneath the canopy (2000–5000 Bq kg^{-1} in 2012 and 1000–2000 Bq kg^{-1} in 2013; Fig. 14.3). In contrast, no significant vertical patterns of ^{137}Cs concentrations was found in the bark of *Q. serrate* (10,000–18,000 Bq kg^{-1}), which did not have a canopy of leaves when the radiocesium was deposited in March 2011. This suggests that aerosols containing radiocesium deposited and adhered to the whole surface of tree trunks. The bark ^{137}Cs concentration decreased to 5000–10,000 Bq kg^{-1} except for the top part of the canopy in 2013. For both the evergreen conifer and deciduous trees, the decrease in ^{137}Cs concentration between 2012 and 2013 implies that the attached radiocesium on the trunk surfaces was leached out by rainwater, and was translocated into other part of the body. Removed with bark abrasion was also possible.

The ^{137}Cs concentration in *C. japonica* wood was 100–210 Bq kg^{-1} for sap wood and 10–220 Bq kg^{-1} for heartwood, and that of *Q. serrate* was 60–-120 Bq kg^{-1} for sap wood and 20–80 Bq kg^{-1} for heartwood (Fig. 14.4). For the coniferous species *C. japonica*, the difference in ^{137}Cs concentrations was small between sapwood and heartwood, indicating that the translocation and/or dispersion occurred quickly within the wood.

While the ^{137}Cs concentration in *C. japonica* wood decreased between 2012 and 2013, the concentration increased for *Q. serrate* wood. This suggests that translocation from the bark, and probably from surface soils including litters, through roots to sapwood and heartwood might be delayed in *Q. serrate* compared to *C. japonica*, or might be greater in *Q. serrate* than *C. japonica*.

All of the results indicate that ^{137}Cs moved actively via nutrient transportation mechanisms and abiotic dispersion into the tree body. In addition to the movement

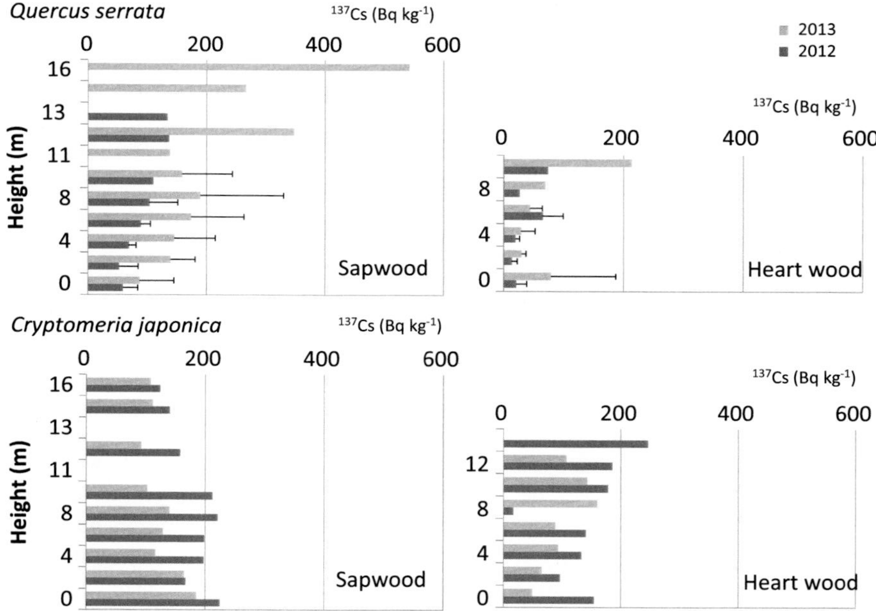

Fig. 14.4 The ^{137}Cs concentrations in sapwood and heartwood of *Quecus serrate* and *Cryptomeria japonica*. Samples were collected from whole tree sampling conducted in early November 2012 and 2013

Table 14.1 Annual average ^{137}Cs concentration of throughfall, stemflow and litterfall, and the estimates of annual ^{137}Cs flux by the different paths

	Annual mean ^{137}Cs concentration			^{137}Cs flux		
	DP1	DP2	CP	DP1	DP2	CP
			(Bq L^{-1})			(Bq m^{-2} $year^{-1}$)
Throughfall	3.10	3.01	5.54	3254	1694	3388
Stemflow	4.01	0.97	2.16	458	101	69
			(Bq kg^{-1})			
Litterfall	8068	7464	17,887	2904	2125	7518

The plot codes; DP1, DP2 and CP indicate the deciduous-pine mixed stand #1, the deciduous-pine mixed stand #2 and the cedar plantation, respectively (calculated from the original data in Endo et al. 2015)

from bark (which continued to have high ^{137}Cs concentrations) to sapwood, translocation to the juvenile leaves was clearly evident. It is currently difficult to evaluate the relative contributions to ^{137}Cs in the wood from root uptake and translocation from the bark under field conditions. The root uptake rate of the ^{137}Cs is the most important factor to be evaluated precisely and urgently.

14.3.2 ^{137}Cs Movement from the Canopy to the Forest Floor

The annual ^{137}Cs fluxes from the canopy to the forest floor via throughfall, stem flow, and litter fall were summarized from October 2012 to September 2013 in Table 14.1. The annual ^{137}Cs flux with litter fall was 2.9, 2.1, and 7.5 kBq m^{-2} $year^{-1}$ for the DP1, DP2, and CP plots, respectively. The largest ^{137}Cs flux in the CP plot can be explained by the highest ^{137}Cs concentration in the litter from *C. japonica*.

The annual ^{137}Cs fluxes for the sum of throughfall and stem flow were 3.7, 1.8, and 4.0 kBq m^{-2} $year^{-1}$ for the DP1, DP2, and CP plots, respectively. The rate values were in the same order as those for litter fall. The total ^{137}Cs flux by these pathways were 6.6, 3.9, 11.5 kBq m^{-2} $year^{-1}$ for the DP1, DP2, and CP plots, respectively.

Each of these fluxes changed temporally depending upon phenology, seasonal variation of precipitation rate, and frequency of storm events (the data were not shown. See Endo et al. 2015). The ^{137}Cs flux with litter fall was largest in late fall (late October to mid-November) while that with throughfall and stem flow was high in the middle of summer (July to September) when the precipitation rate was greatest. The ^{137}Cs movement from the canopy to the forest floor occurred actively through the plant-growing season, even 2 years after the radiocesium was deposited.

The ^{137}Cs availability for plants, algae, and microbes might be different between those attached to litters and those dissolved or suspended in throughfall or stem flow water. In order to understand the current mechanisms of radiocesium cycling between plants and soils, it is important to quantify the bioavailable ^{137}Cs in the forest floor including the litter layer and the surface layer of mineral soils where the activities of roots and microorganisms are high.

14.3.3 ^{137}Cs Discharge from the Forest Catchment

After the radioactive substances were deposited in March 2011, several reports describing radiocesium transportation through the rivers in Fukushima and surrounding areas were published. One report found that 90 % of the ^{137}Cs discharged to the coast through the river were attached to suspended solids (SS), and 60 % of those were discharged during flooding (Yamashiki et al. 2014). The reports all stated the importance of evaluating the quantity and quality of SS in order to quantify ^{137}Cs export from the catchment.

Figure 14.5 represents an example of the short-term temporal change in SS concentration and the ^{137}Cs concentration with the discharge hydrograph during the storm event on October 15, 2013. The SS concentration increased with an increase

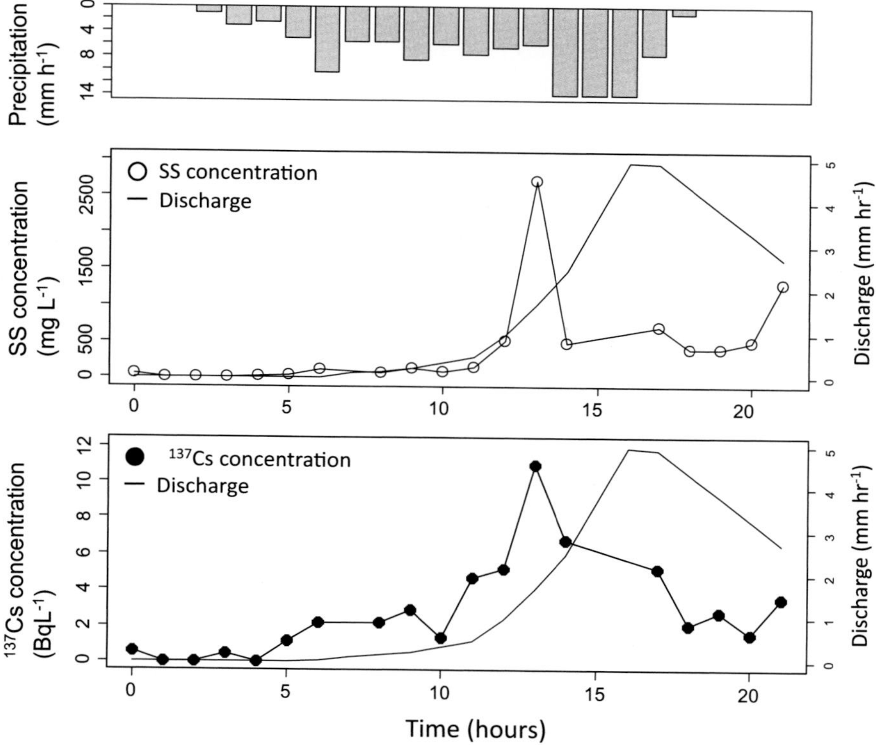

Fig. 14.5 Temporal variations of precipitation, suspended solid (SS) concentration, ^{137}Cs concentration, and river discharge rate on October 15, 2013

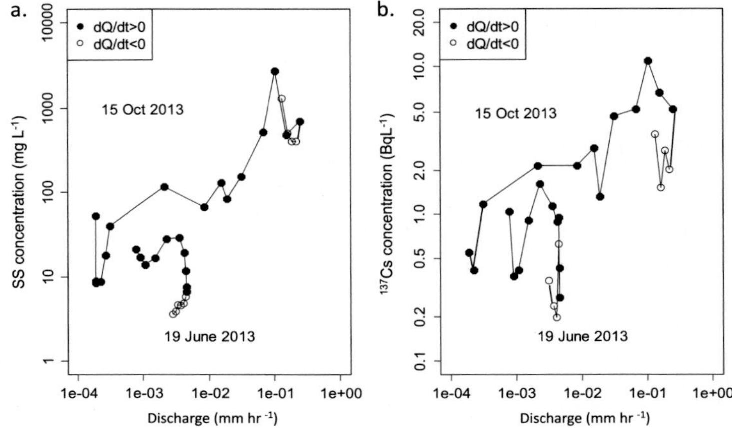

Fig. 14.6 Relationships between (1) discharge and suspended solid (SS) concentration, and (2) discharge and ^{137}Cs concentration. Water samples were collected on June 19 and October 15, 2013. dQ/dt > 0 and dQ/dt < 0 indicate the discharge increasing period and discharge decreasing period, respectively

in river discharge, and began to decrease before the river discharge reached a peak value. The response of ^{137}Cs concentration to the river discharge was similar to that of the SS concentration, evidencing that the ^{137}Cs was transported by the SS.

As shown in Fig. 14.6, the changes of the SS and ^{137}Cs concentrations corresponding to the river discharge rate was different between the period of increasing discharge and decreasing discharge. This suggests that the major source of the discharging ^{137}Cs was the surface soils and litter detritus from the areas near the riverbank. Those materials can be easily washed out under high flow conditions during a storm.

In the case of the event on October 15, 2013, the decrease in SS and ^{137}Cs concentrations before the peak discharge moment suggested that the limitation in SS and ^{137}Cs sources occurred during the increasing discharge period (Figs. 14.5 and 14.6).

The mechanism of washing ^{137}Cs from the riverbanks might be activated by storm events that occur several to several ten times per year. This means simultaneously that the supply of ^{137}Cs onto the forest floor of this area (riverbanks) has occurred continuously. The substantial mechanism of this ^{137}Cs supply was litter fall, throughfall, and stem flow as we explained earlier in this chapter.

The rating function was established from those relationships between the discharge rate and the ^{137}Cs concentration. The annual ^{137}Cs discharge (1-year accumulated value) using this function was estimated to be 330 Bq m^{-2} year^{-1} for the period from August 31, 2012 to August 30, 2013, while 670 Bq m^{-2} year^{-1} for the period from October 23, 2012 to October 22, 2013. This difference was attributed to the large storm event in the mid October 2013, which discharged 227 Bq m^{-2} of ^{137}Cs in a single flood event. This emphasizes the importance of quantifying the influence of storm events on the amount of ^{137}Cs discharged.

The annual ^{137}Cs discharge through the river was one order of magnitude smaller than the annual total flux of ^{137}Cs movement from the canopy to the forest floor. This means that the source limitation of the ^{137}Cs has not occurred for the discharge process on a yearly basis.

However, the estimated amount of the ^{137}Cs deposited initially in March 2011 was 100–300 kBq m^{-2} (Ministry of Education, Culture, Sport, Science and Technology 2013). This is three orders of magnitude larger than the quantity of ^{137}Cs discharged through the river, indicating that in this forest ecosystem the proportion of the ^{137}Cs pool that is discharged through the river is significantly less than the proportion lost by radioactive decay.

14.3.4 ^{137}Cs Dispersion Through Food Web

As mentioned above, the ^{137}Cs has already been in the wood of the dominant trees. It is possible that ^{137}Cs has begun to circulate among plants and soils. It could be inferred that the food web has already received the ^{137}Cs dispersion to some extent from the primary producers.

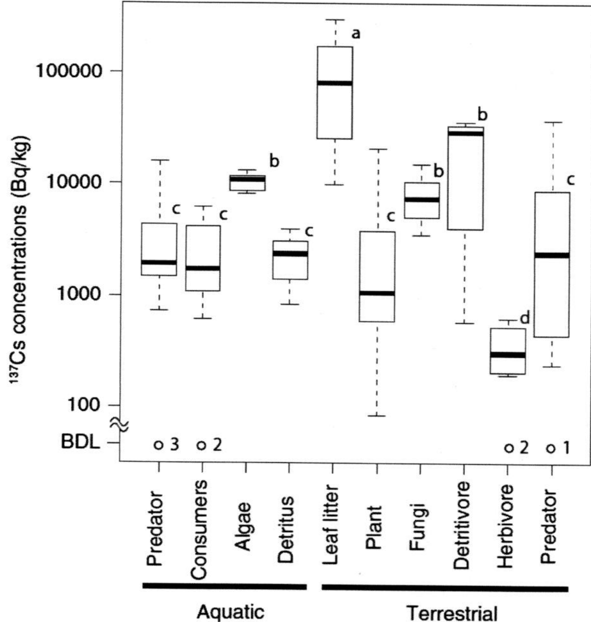

Fig. 14.7 ^{137}Cs concentrations in each functional group. *Different letters beside each box* indicate differences in ^{137}Cs concentrations based on the grouping of functional groups with the model selection using glm. Data below the detection limit were excluded from the analysis. Numbers beside the BDL (below the detection level) symbols show the number of specimens BDL (Murakami et al. 2014)

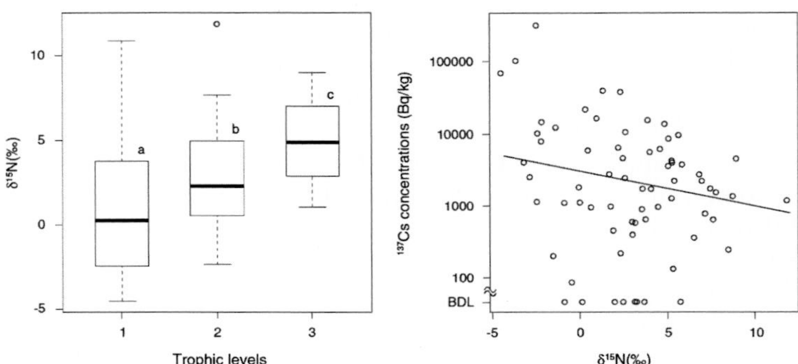

Fig. 14.8 Relationship between δ^{15}N values and ^{137}Cs concentrations in each sample. Equation: $\log(^{137}\text{Cs}) = -15.7(\delta^{15}\text{N}) + 1392.3$. Data below the detection limit were excluded from the analysis (Murakami et al. 2014)

Figure 14.7 shows the observed ^{137}Cs concentrations in aquatic and terrestrial organisms classified into functional feeding groups. The ^{137}Cs concentrations in leaf litters, fungi, detritivores, and predators were significantly higher than those in living plant leaves and their herbivores. The large accumulation of ^{137}Cs in litters and litter detritus indicates that the detrital food chains in the terrestrial community are major pathways that substantially transfer ^{137}Cs. In the aquatic community, the ^{137}Cs concentrations in leaf detritus and algae (the base foods) were between those of living leaves and litters, and reflected on those of organisms at higher trophic levels.

As proxies for the relative heights of the trophic levels (Fig. 14.8a), the δ^{15}N values tended to increase as the ^{137}Cs concentrations in organisms decreased (Fig. 14.8b). This decrease in ^{137}Cs concentrations through trophic interactions, suggests biological dilution and not accumulation of ^{137}Cs.

14.4 Summary and the Future Directions

Biogeochemical storage, cycling, distribution, and movement of radiocesium in the forested ecosystem were assessed for about 3.5 years after the Fukushima Daiichi Nuclear Power Plant accident. The most important finding was that the ^{137}Cs movement within the forested ecosystem was still active, and the ^{137}Cs discharge through the river was significantly small compared to the deposited amount of ^{137}Cs in this area.

The majority of the ^{137}Cs pool was in the litter layer and the shallow soil horizons. The attached ^{137}Cs on the tree canopies and trunks was gradually removed and significantly transferred to the forest floor. Simultaneously, ^{137}Cs moved into the tree bodies from the surfaces and probably also via root uptake from the litter layer and soils, while ^{137}Cs had been transported into the new leaves since the early phase of monitoring. Although the ^{137}Cs transformation from the tree canopies to the forest floor was still active in 2012 and 2013, it can be inferred that the transferring flux will decrease gradually with the removal of the attached ^{137}Cs. It can also be hypothesized that the "equiribulium" will be made between the downward ^{137}Cs transferring flux and the uptake flux by plants. This internal cycling between soils and plants will play important role to retain the relatively mobile forms of ^{137}Cs in the forest ecosystem, even though certain portion of the ^{137}Cs in the litter layer moves into the mineral soil horizons, and is fixed strongly by the cray minerals.

Detrital food chains, based on leaf litter detritus as the primary carbon source, have introduced and dispersed ^{137}Cs through the food web, even to the top predators. However, no evidence of bioaccumulation was detected from the present observations.

An urgent question to address is when and how the ^{137}Cs diffuses through the grazing food chain and the detritus food chain, and how rapidly does this process occur? The key aims are not only to predict this phenomenon, but also to precisely understand the circulation and redistributions among the system of the soil-plant nutrient cycle, the mechanisms of ^{137}Cs release from litter and soil organic matter, as well as the ^{137}Cs-absorbing behavior of plants.

As we reported in this chapter, radiocesium movement within the forest ecosystem is still active and has not reached a steady state. In order to determine the long-term (e.g., equivalent to the ^{137}Cs half-life) influences of radiocesium existence in the forest ecosystem, we still need to carefully monitor the succession of changes in radiocesium distributions.

Acknowledgments All data were collected during a research project supported by a grant (24248027) for scientific research from the Ministry of Education, Culture, Sports, Science and Technology and by a grant for river management research (FY2014) from the River Foundation (Kasen Zaidan, Japan). The authors would like to thank Mr. Chonosuke Watanabe for his kind support during field work, and also thank Dr. Takashi Nakamura of University of Yamanashi for his technical support with stable isotope analysis.

References

Bunzl K, Schimmack W, Kreutzer K, Schierl R (1989) Interception and retention of chernobyl-derived ^{134}Cs, ^{137}Cs and ^{106}Ru in a spruce stand. Sci Total Environ 78:77–87

Endo I, Ohte N, Iseda K, Tanoi K, Hirose A, Kobayashi NI, Murakami M, Tokuchi N, Ohashi M (2015) Estimation of radioactive 137-cesium transportation by litterfall, stemflow and throughfall in the forests of Fukushima. J Environ Radioact 149:176–185

Fukuyama T, Takenaka C, Onda Y (2005) ^{137}Cs loss via soil erosion from a mountainous headwater catchment in central Japan. Sci Total Environ 350:238–247

Hashimoto S, Ugawa S, Nanko K, Shichi K (2012) The total amounts of radioactively contaminated materials in forests in Fukushima, Japan. Sci Rep 2:416

Hashimoto S, Matsuura T, Nanko K, Linkov I, Shaw G, Kaneko S (2013) Predicted spatio-temporal dynamics of radiocesium deposited onto forests following the Fukushima nuclear accident. Sci Rep 3:2564

Hisadome K, Onda Y, Kawamori A, Kato H (2013) Migration of radiocaesium with litterfall in hardwood-Japanese red pine mixed forest and sugi plantation. J Jpn For Soc 95:267–274 (In Japanese with English abstract)

Kato H, Onda Y, Gomi T (2012) Interception of the Fukushima reactor accident-derived ^{137}Cs, ^{134}Cs and ^{131}I by coniferous forest canopies. Geophys Res Lett 39:L20403

Kinnersley RP, Goddard AJH, Minski MJ, Shaw G (1997) Interception of caesium-contaminated rain by vegetation. Atmos Environ 31:1137–1145

Kitchings T, Digregorio D, Van Voris P (1976) A review of ecological parameters in vertebrate food chains. In: Proceedings of the fourth national symposium on radioecology. Ecological Society of America, Oregon State University, Corvallis, pp 304–313

Kruyts N, Delvaux B (2002) Soil organic horizons as a major source for radiocesium biorecycling in forest ecosystems. J Environ Radioact 58:175–190

Ministry of Education, Culture, Sport, Science and Technology and Ministry of Ministry of Agriculture, Forestry and Fisheries (2012) Study report on distribution of radio active substances emitted by the accident of the Fukushima Daiichi Nuclear Power Plant, Tokyo

Ministry of Education, Culture, Sport, Science and Technology (2013) Distribution map of radiation dose rate. http://ramap.jmc.or.jp/map/mapdf/pdf/air/20131119/dr/5640-C.pdf. Accessed 3 Feb 2015

Murakami M, Ohte N, Suzuki T, Ishii N, Igarashi Y, Tanoi K (2014) Biological proliferation of cesium-137 through the detrital food chain in a forest ecosystem in Japan. Sci Rep 4:3599

Ohte N, Murakami M, Suzuki T, Iseda K, Tanoi K, Ishii N (2012) Diffusion and export dynamics of ^{137}Cs deposited on the forested area in Fukushima after the nuclear power plant accident in March 2011: preliminary results. In: Proceedings of the international symposium on environmental monitoring and dose estimation of residents after accident of TEPCO's Fukushima Daiichi nuclear power stations. KUR research program for scientific basis of nuclear safety, Kyoto, pp 25–32

Ohte N, Murakami M, Iseda K, Tanoi K, Ishii N (2013) Diffusion and transportation dynamics of ^{137}Cs deposited on the forested area in Fukushima after the nuclear power plant accident in March 2011. In: Nakanishi T, Tanoi K (eds) Agricultural implications of the Fukushima nuclear accident. Springer, New York, pp 177–186

Rowan DJ, Rasmussen JB (1994) Bioaccumulation of radiocesium by fish: the influence of physicochemical factors and trophic structure. Can J Fish Aquat Sci 51:2388–2410

Schimmack W, Förster H, Bunzl K, Kreutzer K (1993) Deposition of radiocesium to the soil by stemflow, throughfall and leaf-fall from beech trees. Radiat Environ Biophys 32:137–150

Soil Survey Staff (2014) Keys to soil taxonomy, 12th edn. USDA-Natural Resources Conservation Service, Washington, DC

Wakiyama Y, Onda Y, Mizugaki S, Asai H, Hiramatsu S (2010) Soil erosion rates on forested mountain hillslopes estimated using ^{137}Cs and $^{210}Pb_{ex}$. Geoderma 159:39–52

Wang WX, Ke C, Yu KN, Lam PKS (2000) Modeling radiocesium bioaccumulation in a marine food chain. Mar Ecol Prog Ser 208:41–50

Yamashiki Y, Onda Y, Smith HG, Blake WH, Wakahara T, Igarashi Y, Matsuura Y, Yoshimura K (2014) Initial flux of sediment-associated radiocesium to the ocean from the largest river impacted by Fukushima Daiichi Nuclear Power Plant. Sci Rep 4:3714

Chapter 15
Reduction of Air Radiation Dose by Ponding Paddy Fields

Naritaka Kubo, Toshiaki Iida, and Masaru Mizoguchi

Abstract Radioactive cesium (Cs) released by nuclear accidents is sorbed and fixed onto soil surfaces, which then radiate strong gamma rays (γ-rays). Decontamination around dwelling areas is now eagerly being implemented but more efforts are necessary to reduce the air radiation dose. Paddy field ponding, from the viewpoint of cost-effectiveness, is considered to be an effective practice for reducing the air radiation dose in the environment. In this study, field experiments were conducted at Sasu and Komiya regions in Iitate Village to verify the effectiveness of paddy field ponding, and numerical experiments were also conducted using the formula for uncollided γ-ray fluxes passing through the shield material.

It was found that the a ponding water depth of 20–25 cm can drastically reduce the number of γ-ray photons emitted from the paddy fields, and the reduction in radiation dose was related to water depth. However, some differences were also observed between field and numerical experiments. The numerical calculation showed that the radiation dose decreased exponentially when the depth increased; however, field experiments showed a linear decrease. The cause might be the build-up effect caused by Compton scattering, but the details are unclear. It is necessary to explain these differences before ponding becomes a useful practice.

Keywords Decontamination • Gamma ray • Paddy field ponding • Puddling • Radiation shielding • Half-value thickness

Abbreviation

Cs	Cesium
γ-ray	Gamma ray
Bq	Becquerel
Sv	Sievert

N. Kubo (✉) • T. Iida • M. Mizoguchi
Graduate School of Agricultural and Life Sciences, The University of Tokyo, 1-1-1, Yayoi, Bunkyo-ku, Tokyo 113-8657, Japan
e-mail: anakubo@mail.ecc.u-tokyo.ac.jp

T.M. Nakanishi, K. Tanoi (eds.), *Agricultural Implications of the Fukushima Nuclear Accident*, DOI 10.1007/978-4-431-55828-6_15

15.1 Introduction

Three and a half years have passed since the Great East Japan earthquake (March 11, 2011), and earnest decontamination works are now being implemented around the dwelling areas of IITATE village. Radioactive substances are surely reduced by the decontamination works and are also gradually diminishing through runoff and decay. In the near future, the evacuation order will be lifted at "zones in preparation for the lifting of the evacuation order" successively when the air radioactive dose becomes sufficiently low (Minyu-net 2014). However, more efforts are necessary to reduce the radioactivity before the evacuated villagers return. Among some of the measures is "paddy field paddling and ponding," which is considered to be feasible in terms of cost-effectiveness and farmland conservation.

Gamma rays (γ-rays), which are generated when radioactive Cesium (Cs) decays, are a form of electromagnetic radiation like visible light; they behave just like particles and are emitted in every direction (Tazaki 2011). In the case of paddy fields, Cs fixes onto the soil surface (Shiozawa et al. 2011), and because the field surface is flat, the soil can be a major source of γ-rays if there is no obstructive shielding. However, if the paddy fields are ponded, the γ-rays coming from the paddy field can be reduced considerably. Thus, even if the living space is close to the paddy field, the γ-ray photons are substantially reduced because the γ-rays heading toward the living space have a low elevation angle and they must cross the ponded water layer obliquely with long path length.

Most Japanese paddy fields are irrigated unlike upland fields, and they are equipped with irrigation facilities to supply them with water. Some facilities were damaged by the earthquake, but minor repairs and maintenance can recover their functions. Only supplying water to the paddy field, however, cannot attain maintaining ponding depth; they need puddling and border coating. Without these practices, ponded water is rapidly lost by vertical and horizontal percolation (Lee et al. 2003). Puddling disperses Cs within the plow layer, and the γ-ray dose can be reduced by soil and saturated soil water. The puddling also prevents the dust, which sorbs Cs, from being blown up. Besides, paddy field ponding is effective for preventing the growth of weeds and invasion by wild animals like wild boars and monkeys. If the soil-to-rice transfer of radioactive cesium is permissible, then rice cultivation in a deeply ponded paddy field will be a big step for reconstructing the village.

In this paper, the effects of paddy field ponding will be verified through field experiments and numerical calculations, and problems will be discussed relating to field experiments and field application.

15.2 Theoretical Consideration

Firstly, the dangers of non-ponded paddy fields will be proven by a simple calculation. Let's consider three radiation sources located at points A($-\varepsilon$, 0), O(0, 0), and B(ε, 0) on the x-axis as shown in Fig. 15.1. Point P(x, y) is remote from O by the distance r, and r is assumed to be much longer than the interval ε ($r \gg \varepsilon$).

The changes of the γ-ray intensity at point P and the visual angle ω ($\angle$APB) are examined when the elevation angle, θ changes. As the γ-rays are emitted in every direction from the source, the intensity is inversely proportional to the square of the distance from the source. The square of the distance from A to P, and the reciprocal of the square can be approximated as follows.

$$\overline{AP}^2 = (x+\varepsilon)^2 + y^2 = r^2\left(1 + 2\frac{x\varepsilon}{r^2} + \frac{\varepsilon^2}{r^2}\right) \cong r^2\left(1 + 2\frac{x\varepsilon}{r^2}\right)$$

$$\frac{1}{\overline{AP}^2} \cong \frac{1}{r^2}\left(1 - 2\frac{x\varepsilon}{r^2}\right)$$

Therefore, as shown below, the total intensity of the γ-rays from three sources is about three times that from only one source at O regardless of the elevation angle, θ.

$$\frac{1}{\overline{AP}^2} + \frac{1}{\overline{OP}^2} + \frac{1}{\overline{BP}^2} \cong \frac{1}{r^2}\left(1 - 2\frac{x\varepsilon}{r^2}\right) + \frac{1}{r^2} + \frac{1}{r^2}\left(1 + 2\frac{x\varepsilon}{r^2}\right) = \frac{3}{r^2}$$

This means that the radiation dose at P is proportional to the number of sources and does not depend on the elevation angle, θ.

On the other hand, when $r \gg \varepsilon$, three lines AP, OP, and BP become almost parallel and the following relationship can be obtained for the visual angle ω, as shown in Fig. 15.2.

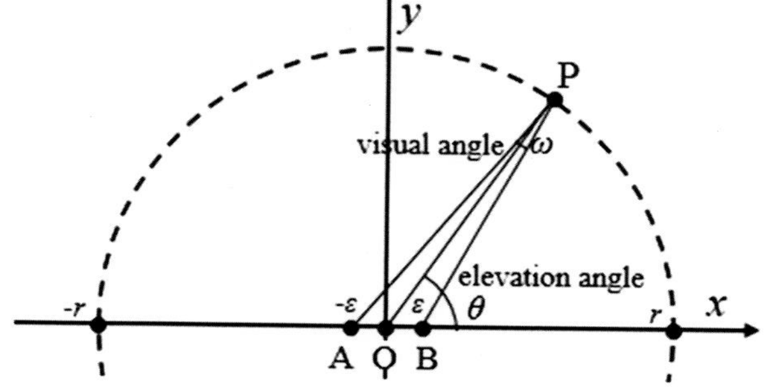

Fig. 15.1 Three radiation point sources at A, O, and B on the plane and a receiver at P

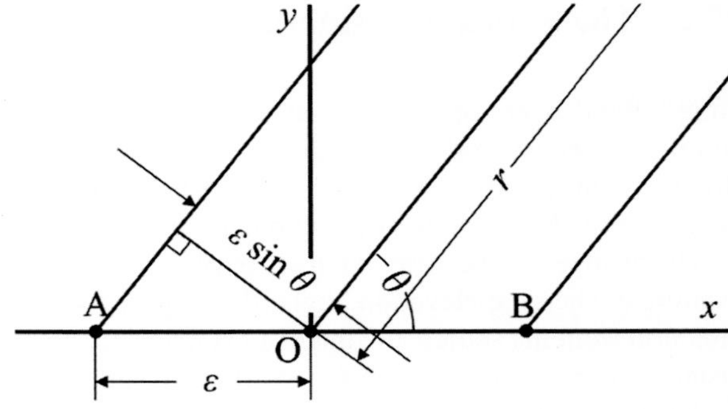

Fig. 15.2 Visual angle ω vs. elevation angle θ when distance r is much longer than interval ε

$$\sin(\omega/2) \cong \varepsilon \sin\theta \div r$$

Furthermore, as $\sin(\omega/2) \cong \omega/2$ when ω is small, the visual angle ω is approximated by the following equation.

$$\omega \cong 2(\varepsilon/r)\sin\theta$$

The visual angle takes a maximum value of $2(\varepsilon/r)$ when $\theta = \pi/2$, and is zero when $\theta = 0$. As the total intensity of the γ-ray is constant regardless of θ, the intensity per unit visual angle becomes very strong (bright) when the visual angle ω, relative to the elevation angle θ, becomes small. The above considerations are realized when the paddy field is undisturbed, because the Cs settles from the air onto the soil surface, and the paddy field surface is artificially flat. Consequently, the undisturbed paddy fields are very dangerous for people staying in the course of the γ-rays emitted with a small elevation angle.

Secondly, the shield effectiveness of soil or water against the γ-rays will be examined. Gamma radiation is an electromagnetic wave, but it behaves as a photon and is mainly attenuated by electron to γ-ray interactions when it passes through a substance. The number of photons is attenuated $e^{-\mu d}$ times when they pass through a substance having a thickness of d. This μ is called the linear attenuation coefficient, and μ_m ($= \mu/\rho$, where ρ is density) is called the mass attenuation coefficient (Tazaki 2011). The value of μ_m is almost constant regardless of the kind of substance, and therefore the value of μ is nearly proportional to the density of the substance. The thickness of $d_{0.5}$, through which the number of photons is halved, is called the half-value thickness. The values of $d_{0.5}$ for water, air, and soil are 8.1 cm, 70 m, and about 5 cm, respectively, for γ-rays emitted from ^{137}Cs (Fujiwara 2011).

Figure 15.3 shows that the pass length is $d/sin\theta$ when the thickness is d and the elevation angle is θ. The reduction ratio of photons is calculated by the following equation.

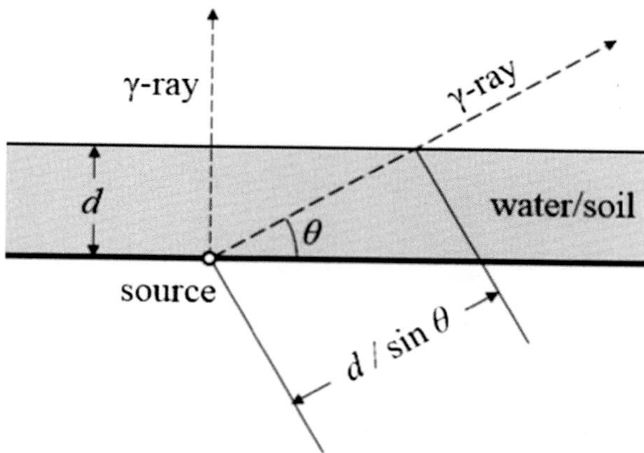

Fig. 15.3 Travel length of the γ-ray in the medium of water or soil

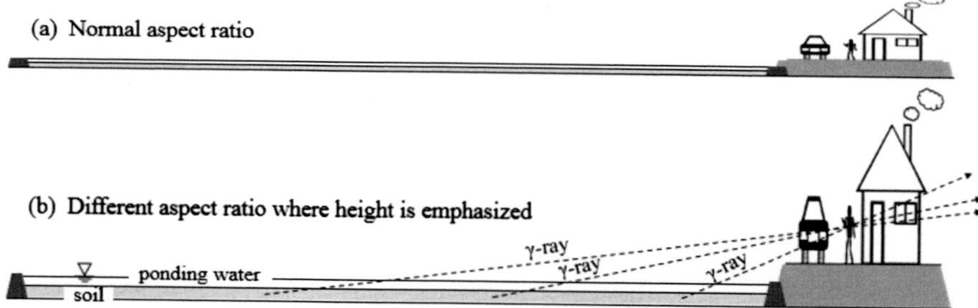

Fig. 15.4 Gamma rays radiated from the paddy field to the living space

$$0.5^{\frac{d}{d_{0.5}} \times \frac{1}{sin\theta}} = \left(0.5^{\frac{d}{d_{0.5}}}\right)^{\frac{1}{sin\theta}}$$

Contrary to this case, the shield effectiveness of water/soil is enormous when the elevation angle is small, and such shield effect is superior to the amplification effect due to the low elevation angle. Specifically, if the thickness d is equal to the half-value thickness $d_{0.5}$, then the reduction ratios of photons are 1/4, 1/15, 1/55, and 1/3000, and the brightness ratios are 1/2, 1/4, 1/10, and 1/250 when the elevation angles are 30°,15°, 10°, and 5°, respectively.

The above consideration shows that paddy fields are no longer dangerous if the radioactive substances are mixed with soil by plowing, the mixed soil is then saturated with water by puddling, and the paddy fields are kept deeply ponded. All of these measures can considerably reduce the number of photons that would otherwise be emitted to the air. They can be especially effective for enabling daily living spaces, which are in the course of the γ-rays with a low elevation angle. Figure 15.4 shows the γ-ray radiation from paddy fields to living spaces: (**a**) the aspect ratio is normal and (**b**) the aspect ratio is different to emphasize the height.

It can be seen from Fig. 15.4 that the actual elevation angle of the γ-ray course is very low, and the pass length in soil/water becomes very long, which will contribute to attenuation of the γ-ray intensity.

15.3 Field Experiments

Two sections in IITATE village, Sasu and Komiya, were selected for the field experiments to examine the effects of ponding paddy fields on the reduction of air radiation. The experiment at Sasu was preliminary and the observation period was less than 1 month from the 13th of October 2012 to the 10th of November 2012. The main experiment was at Komiya, where the observation period was 4 months from the 3rd of August 2013 to 3rd of December 2013.

15.3.1 Field Experiment at Sasu

This section was designated as "zone in preparation for the lifting of the evacuation order," and the radiation dose was relatively low. The experimental paddy field was situated at lat. 37°44′ 15″ and long. 140°43′ 44″, and was located at the bottom of a hill surrounded by forests extending to the west and east.

Figure 15.5 shows the configuration of the paddy fields and locations of observation equipment for the γ-ray intensity (γ-ray) and ponding depth (WL). The point A is located at the corner of the paddy field, and the point B is at the center of the paddy field but is 1 m outside the border. Water was taken from an irrigation ditch and was drained to the natural stream nearby. Only one paddy field was supplied with water, but the others were not. However, some partial ponding was observed in the right-side neighboring paddy field, which received some percolating water that seeped out from the ridge. A Geiger Müller counter installed 1 m above the paddy field surface measured the γ-ray intensity. This type of counter, however, cannot distinguish decaying radioactive elements, and therefore it was difficult to accurately measure radioactive doses.

The amount of ^{134}Cs and ^{137}Cs decreases by natural decay over time. About 2.8 % of ^{134}Cs and 0.2 % of ^{137}Cs decreased in 1 month, but these decreases were not compensated for in the experiment at Sasu. Figure 15.6a shows the time series of the γ-ray counting per hour at points A and B. The counts represent the daily mean because it fluctuated intensely. Figure 15.6b shows the time series of ponding water levels, the reference of which is the mean altitude of the paddy field surface. The elevation of point C was about 6 cm lower than the mean value because the field surface was slightly inclined to the north side, and water depth $h = 0$ cm corresponds to $WL = -6$ cm.

Figure 15.7a shows the relationship between the ponding water level and the γ-ray counts at points A and B. The radiation counts were observed to decrease by

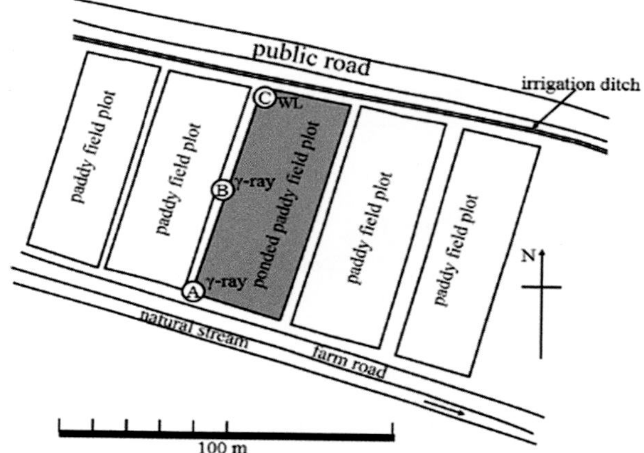

Fig. 15.5 Experimental paddy field in Sasu. Section and locations of observation stations for γ-rays and water levels

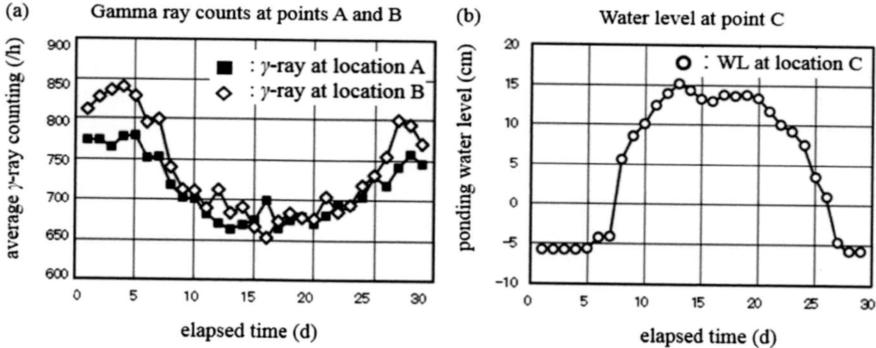

Fig. 15.6 Field observations of γ-ray radiation and water level

4.57 counts at point A and by 6.79 counts at point B, which corresponded to a rise in water level of 1 cm. Figure 15.7b shows the correlation of concurrent radiation counts at points A and B, and shows that the decreasing rate at B was 1.43 times larger than that at A.

Theoretically, the uncollided γ-ray flux can be calculated against the ponding depth; however, it does not directly correspond to the effective dose because it does not consider scattering. The flux I, the number of γ-ray photons, which pass per unit time through an orbicular head having a unit cross section, can be calculated using the following equation.

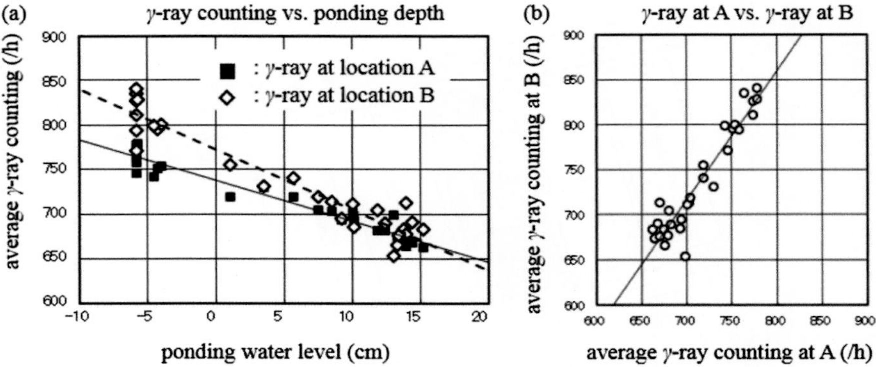

Fig. 15.7 Relationship between γ- ray counts and ponding depth at points A and B

$$I = \frac{p}{4\pi} \iint_S \frac{(0.5)^{\sqrt{r^2+h^2} \times \left(\frac{h-d}{h\, d_{0.5a}} + \frac{d}{h\, d_{0.5w}} \right)}}{r^2 + h^2} ds = \frac{p}{4\pi} \alpha \tag{15.1}$$

where, p is the number of γ-ray photons emitted from unit paddy field area in unit time, S is the area of the target paddy field, r is the horizontal distance from the observation point to an element having small area ds, h is the height of the γ-ray counter, d is the ponding depth, $d_{0.5\mathrm{a}}$ is the half-value thickness of air, and $d_{0.5\mathrm{w}}$ is the half-value thickness of water.

The integral α in the above equation depends on ponding depth d, height h, and paddy field shape, but the latter two do not change; therefore α is a function of d. In the case where the radiation intensity, p, depends on location and the elapsed time t after the nuclear accident, and if the value for p is assumed to be same in each paddy field, then p is only a function of t. The integrals α_{A} and α_{B} are calculated numerically; the target paddy field of 80 m × 25 m is divided into squares of 0.5 m × 0.5 m, and evenly distributed radioactive substances are assumed to exist at the center of each square, as shown in Fig. 15.8.

Figure 15.9 shows the numerical results of integrals α_{A} and α_{B}. It is difficult to directly compare the measured and calculated results in Figs. 15.7 and 15.9; however, several differences and similarities were observed with regard to the shielding effect of ponding. Firstly, for the reduction pattern of the γ-rays against the ponding depth, the γ-ray counts decreased linearly in Fig. 15.7, but the integral α decreased exponentially in Fig. 15.9.

The effect may be attributed to Compton scattering. The γ-ray counts were measured by using a Geiger Müller counter, which counts γ-rays scattered by the Compton effect as well as uncollided γ-rays. In such cases, the γ-ray flux is built up and "the exponential relation" in Fig. 15.9 may approach to "the linear relation" in Fig. 15.7. Secondly, regarding the reduction rates at points A and B, those at point A were 1.4–1.5 times larger than those at point B. This relationship is seen commonly

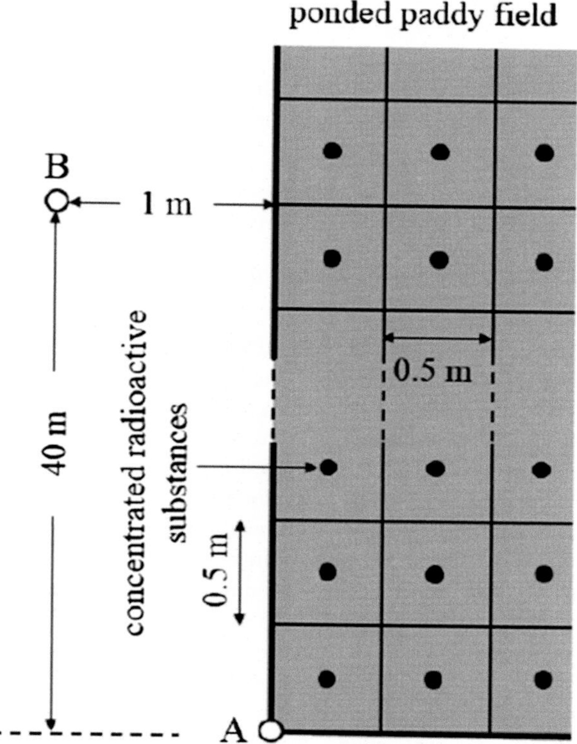

Fig. 15.8 Paddy field and observation points A and B for numerical calculation of uncollided γ-ray flux. Paddy field is divided into *squares* of 0.5 m × 0.5 m and distributing radioactive substances are assumed to be concentrated at the center of each *square*

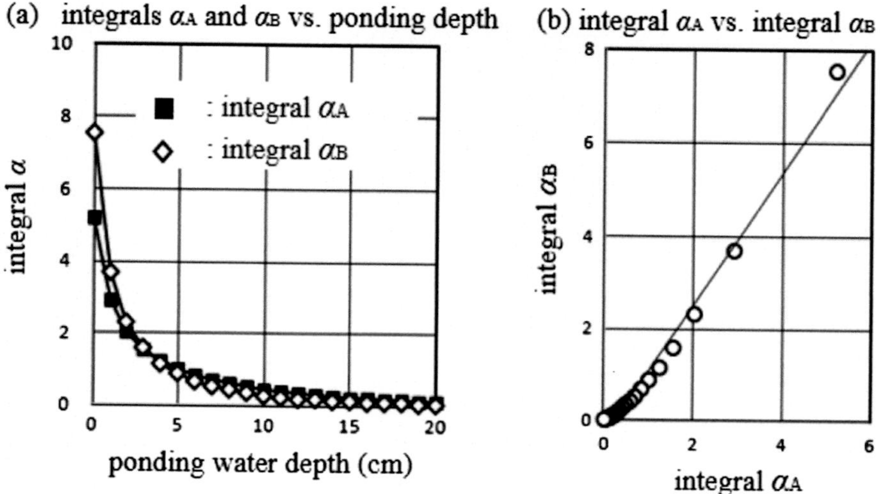

Fig. 15.9 Relationship between integral α and ponding depth at A and B

in the measured and calculated results presented in Figs. 15.7 and 15.9. Theoretically, the reduction rate at B was 2.0 times that at A, if the sensor was set on the border of the paddy field, but actually it was set 1 m outside the border. In any case, if the ponding depth is kept at about 20 cm, then the uncollided γ-ray flux is reduced to 2 % of the original flux intensity, and the γ-ray reduction effects by the water ponding are confirmed to be significant.

15.3.2 Field Experiment at Komiya

This section is designated as a "restricted residential area," and the experimental paddy fields are situated at lat. 37°37′ 31″ and long. 140°46′ 37″ and are surrounded by forests. Figure 15.10 shows the configuration of the paddy fields and the locations of observation equipment for the γ-ray intensity (γ-ray) and ponding depth (WL). Two paddy field plots were used for the experiment. Ponding depths were measured at point F for the left side plot and at point G for the right side plot using U20 water level loggers (Onset Computer Corporation).

The radiation doses were measured at points D and E using GPSGMC-002-TUV loggers (SERIALGAMES Inc.). Point D is on a road, and on both sides of the road are paddy fields. The detailed location is 0.5 and 2.5 m from the left side and right side borders, respectively. Point E is in front of the house and is located 0.5 m outside the paddy field border. Other paddy fields and upland fields are nearby two experimental fields, and forests surround all these fields. These fields and forests are possible sources of γ-ray background, but it is considered to change only mildly

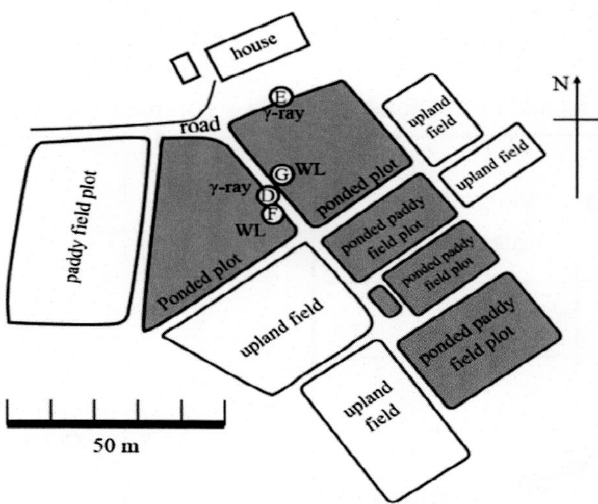

Fig. 15.10 Experimental paddy fields in Komiya section and locations of observation stations for γ-rays and water levels

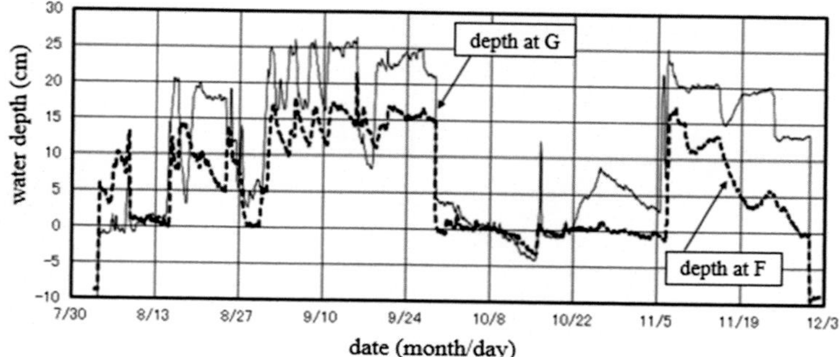

Fig. 15.11 Time series of water depth at F and G

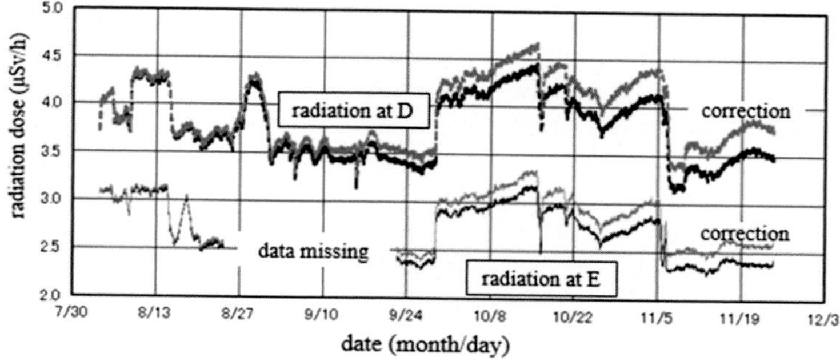

Fig. 15.12 Time series of γ-ray radiation dose at D and E. *Black line*: raw data. *Gray line*: corrected data considering decay

even if it changes seasonally. Figure 15.11 shows the time series of ponding depths at points F and G. The observation period was divided into four for convenience: the August term was from the 3rd to the 31st August with relatively deep ponding depth; the September term was from the 1st to the 28th September with very deep ponding depth; the October term was from the 29th September to the 5th November with very shallow ponding depth; and the November term was from the 6th November to the 3rd December with relatively deep ponding depth.

Figure 15.12 shows the time series of radiation doses at points D and E. At the point E, no data were recorded during the September term because of equipment issues. The observation period at Komiya was 4 months, which is too long to ignore the natural decay, and some kind of correction was needed to compensate for the decrease in radioactivity.

The radiation dose for each day was corrected to the equivalent dose on the 3rd of August assuming the following conditions: the Becquerel (Bq) abundance ratio of ^{134}Cs and ^{137}Cs was 1:1 at the beginning, their half-lives are 2.06 and 30.2 years; the conversion ratio from Bq to Sievert (Sv) is 5.5: 2.1, and the elapsed time is 870 days from the day of the accident to August 3. The black line in Fig. 15.12 shows the before correction dose and the gray line shows the after correction dose. The dose decreased by ~8 % during the 4 months.

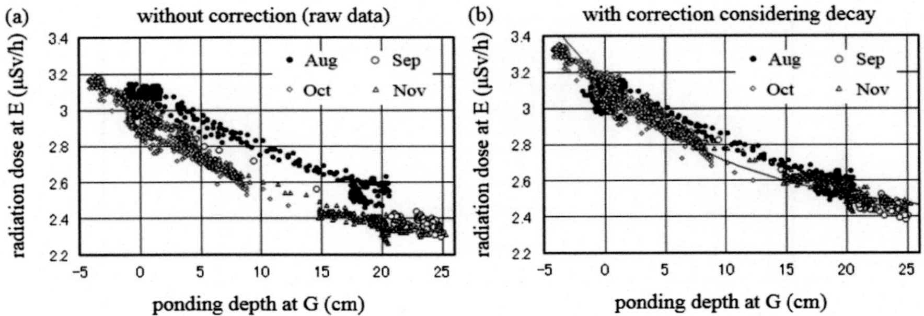

Fig. 15.13 Radiation dose at E (*in front of the house*) vs. ponding depth at G (*right side paddy field*)

Figure 15.13 presents the relationship between the radiation dose at point E and the ponding depth of the right side paddy field (observed at point G). A clear spreading of data can be seen between the early and late data in Fig. 15.13a which uses raw data, but such a spreading of data cannot be seen in Fig. 15.13b which uses corrected data. Considering the above results, our analyses from this point on will use corrected data (values equivalent to those on August 3rd). The radiation dose corresponding to the increase of the ponding depth decreased exponentially, although it decreased slowly. The regression curve of the radiation dose y (μSv/h) may be approximated by an exponential function of depth h (m) as follows.

$$\begin{aligned} \mathrm{y} &= 2.38 + 1.16 \times \mathrm{Exp}(-8.38 \times (h + 0.05)) \\ &= 3.54 - 1.16 \times \{1 - \mathrm{Exp}(-8.38 \times (h + 0.05))\} \end{aligned} \tag{15.2}$$

This equation shows that the radiation dose was 3.54 (μSv/h) if the field was not ponded ($h = -0.05$ m, because of the uneven field surface), but it decreased to 2.38 (μSv/h), which corresponds to background radiation, if the γ-rays from the relevant field were perfectly obstructed.

Figure 15.14 shows the time series of the radiation dose at E; the solid line is observed and the dotted line is estimated from the depth at point G using this regression curve. This regression curve can complement the lack of data, but it tends to overestimate dose values when they are large.

Although point E is scarcely influenced by the far left side paddy field, point D is surely influenced by paddy fields on both sides. Figure 15.15a shows the relationship between the radiation dose at D and the averaged ponding depth at F and G. The plotted points are distributed over a wide range, especially the scattered points marked by Δ which were observed in November when the ponding depths at either side were considerably different. Figure 15.15b shows the relationship between the radiation dose at D and the ponding depth at point F in the left side paddy field; the plotted points are still distributed over a wide range.

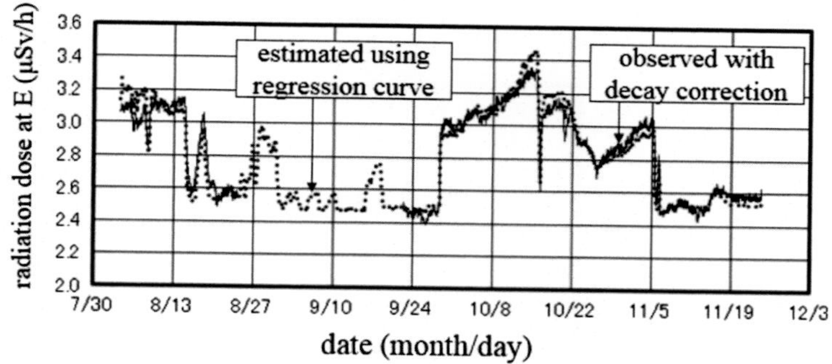

Fig. 15.14 Estimated radiation dose at E using a regression *curve*

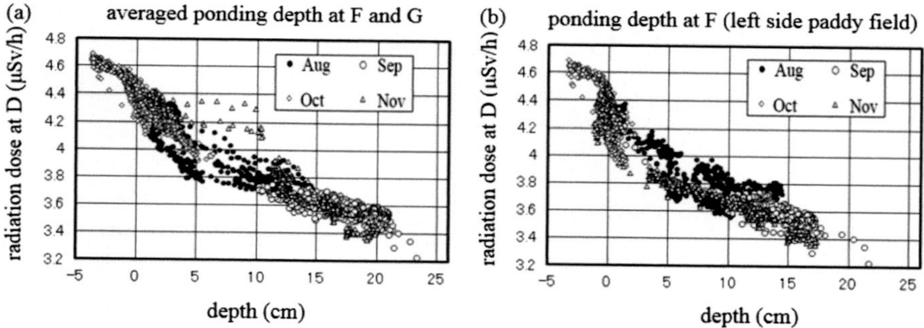

Fig. 15.15 Radiation dose at D (on the road) vs. ponding depth

To make the relationship clear between the radiation dose at D and the ponding depth at F, the influence of the right side paddy field must be removed. Accordingly, a numerical calculation of uncollided γ-ray flux was utilized to estimate the influence of the right side paddy field on radiation doses at points D and F. The shape of the right side paddy field is a distorted rectangle as seen in Fig. 15.10, but it is approximated to a rectangle in Fig. 15.16 with the point D being 2.5 m from the long side border and the point E being 0.5 m from the short side border. As the value of p in Eq. (15.1) was assumed constant, the radiation doses at D and F are proportional to the integrals α_{D} and α_{E}, respectively.

If α_{D0} is the integral at point D when water is not ponded, then the value of $(\alpha_{D0} - \alpha_D)$ is the decrement caused by ponding water, and similarly for point E. This means that the decrement at point D, when the right side paddy field is ponded, is $(\alpha_{D0} - \alpha_D)/(\alpha_{E0} - \alpha_E)$ times the decrement at point E. Figure 15.17 shows the results of numerical calculation of uncollided γ-ray flux; (**a**) shows integrals of α_{D} and α_{E} vs. ponding depth at point G, and (**b**) shows the ratio $(\alpha_{D0} - \alpha_D)/(\alpha_{E0} - \alpha_E)$. As the radiation dose at point E vs. ponding depth at point G was formulated by Eq. (15.2), the reduction effect Δy by ponding can also be formulated by the following equation.

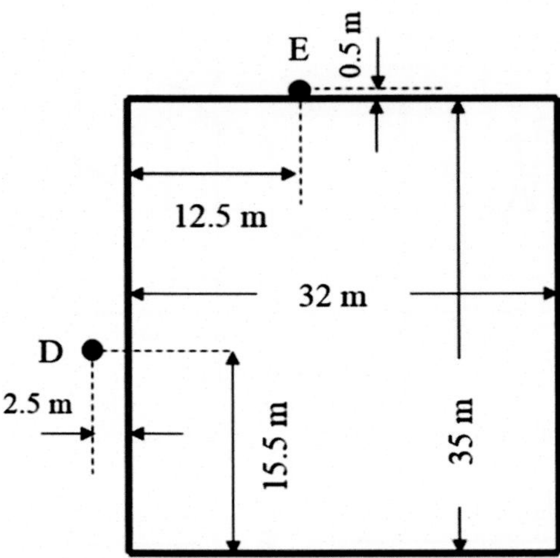

Fig. 15.16 Approximated paddy field for numerical integration. D: Gamma ray observation point on the road. E: Gamma ray observation point in front of the house

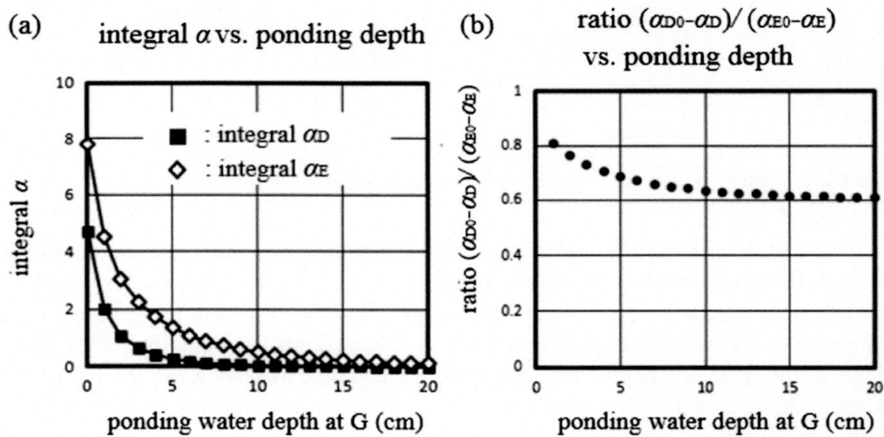

Fig. 15.17 Numerical integral at D and E to compare radiation doses

$$\Delta y = 1.16 \times \{1 - \text{Exp}(-8.38 \times (h + 0.05))\}$$

The radiation dose at point D, when the depth at F is variable but the depth at G is maintained at −0.05 m, can be estimated by adding $\Delta y \times (\alpha_{D0} - \alpha_D)/(\alpha_{E0} - \alpha_E)$ to the observed radiation dose at D.

Figure 15.18 shows the estimated radiation dose at D vs. ponding depth at F in the left side paddy field. Compared to the relationship in Fig. 15.15, the radiation dose is more clearly related with ponding depth. The regression curve for plotted points in Fig. 15.18 can be formulated using an exponential function as follows.

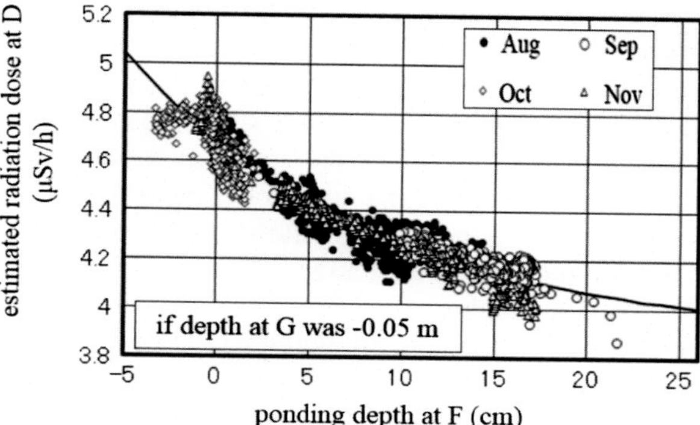

Fig. 15.18 Estimated radiation dose vs. ponding depth at F if ponding depth at G was kept at −0.05 m

$$\begin{aligned} y &= 3.90 + 1.06 \times \text{Exp}(-7.59 \times (h + 0.04)) \\ &= 4.96 - 1.06 \times \{1 - \text{Exp}(-7.59 \times (h + 0.04))\} \end{aligned} \tag{15.3}$$

This equation implies that the radiation dose at point D is 4.96 (μSv/h) if both paddy fields are not ponded, but it will go down to 3.90 (μSv/h) if the ponding depth of the left side paddy field is deep enough. Furthermore, if both paddy fields are deeply ponded, then the radiation dose is reduced to $3.90 - 1.16 \times 0.6 \cong 3.20$ (μSv/h). When the two regression equations of Eqs. (15.2) and (15.3) are compared, the background radiation doses are slightly different, 3.54 (μSv/h) at E and 4.96 (μSv/h) at D; the potential decrements are similar, 1.16 (μSv/h) at E and 1.06 (μSv/h) at D; and the constants of exponent are also similar, −8.38 at E and −7.59 at D. However, in case of the uncollided γ-ray flux shown in Fig. 15.17a, the constant of the exponent ranges from −20 to −30 and is fairly different from those in Eqs. (15.2) and (15.3). This difference may be attributed to Compton scattering, but the details are unclear.

15.4 Conclusions and Remaining Problems

Through numerical calculations and field experiments we confirmed that paddy fields are potential threats to nearby residents. However, the γ-ray flux levels, and thus the severity of such threats, can be considerably reduced by deeply ponding these fields. Moreover, before implementing paddy field ponding, certain problems must be solved. First, the differences between numerical and field experimental results should be quantitatively explained; when the ponding was deepened, the uncollided γ-ray flux decreased exponentially and rapidly, but the radiation dose decreased linearly and gradually. These differences might cause gross errors when the ponding effects are evaluated. Second, water must be deeply and steadily ponded for a long period in the paddy fields. The maximum depth in this field experiment was around 20–25 cm. The following two questions, relating to the maximum depth, should be solved: (1) Is this depth deep enough to shield the γ-rays; and (2) Is it possible to maintain this maximum depth over long periods? According to the numerical calculations, this maximum depth could obstruct nearly all of the uncollided γ-rays, but it is not clear whether this depth can obstruct all effective γ-rays, including the scattered rays. Moreover, it is not easy for an ordinary paddy field to maintain water depths between 20 and 25 cm. Paddy field levees are usually not so high and strong, except for those used for deep ponding irrigation and rainwater storage. They is also not protected against cold weather damage. The heights of levees are usually within the range of 20–30 cm, and the ponding depth is usually no more than 50–60 % of the levee's height for stability. Levee improvement for deep ponding should be examined. The ponded water is continuously lost by evapotranspiration and infiltration, and water must be supplied continuously to maintain ponding. The water supply should not contain Cs, therefore the water quality must be monitored and the standard for water intake should be examined. Paddy field ponding has various merits as mentioned before, but at the same time there remain several problems to be solved before it can be implemented.

Acknowledgment The authors are grateful for the financial support of the Grants-in-Aid for Scientific Research (C), Agriculture Restoration; (Grant No. 25517005). The authors also thank Mr. Muneo Kan-no and Mr. Kin-ichi Ohkubo in IIDATE village for their kind cooperation and sincere help with carrying out the field experiments.

References

Fujiwara T (2011) Calculation of the space dose rate. http://w3.kcua.ac.jp/~fujiwara/nuclear/air_dose.html (in Japanese)

Lee S, Senge M, Ito K, Hayashi H (2003) Influence of direct seeding culture in a well-drained and non-tilled paddy field on water requirement-case study on water requirement of Sunami district. Gifu Prefecture Trans JSIDRE 224:19–26 (in Japanese)

Minyu-net (2014) "difficult-to-return zone," "restricted residence area" and "zone in preparation for the lifting of the evacuation order." http://www.minyu-net.com/osusume/daisinsai/saihen.html (in Japanese)

Shiozawa S, Tanoi K, Nemoto K, Yoshida S, Nishida K, Hashimoto K, Nakanishi T, Nihei N, Ono Y (2011) Vertical concentration profiles of radioactive cesium and convective velocity in soil in a paddy field in Fukushima. Radioisotopes 60:323–328 (in Japanese)

Tazaki S (2011) From Becquerel to Sievert. http://www.gakushuin.ac.jp/~881791/housha/docs/BqToSv.pdf (in Japanese)

Chapter 16
Collaboration Structure for the Resurrection of Iitate Village, Fukushima: A Case Study of a Nonprofitable Organization

Hanae Yokokawa and Masaru Mizoguchi

Abstract Iitate village in Fukushima was evacuated after the Fukushima Daiichi Nuclear Power Plant accident due to the high concentration of radioactive contamination. To revive the serious disaster area, residents, universities, research institutions, experts, and volunteers have collaborated through the nonprofitable organization "Resurrection of Fukushima." The organization is functioning effectively and smoothly based on the members' background, personal connections, and experiences.

Keywords Fukushima Iitate village • Collaboration • Volunteer • Nonprofitable organization (NPO)

16.1 Introduction

Iitate village in Fukushima was evacuated after the Fukushima Daiichi Nuclear Power Plant accident in 2011. High concentrations of radioactive cesium were scattered throughout the village and now the residents from the area are looking to revive the village.

In June 2011, some senior volunteers and researchers started up a NPO (nonprofitable organization) to reconstruct the village in cooperation with the residents. It began with just a few people, but now it involves residents, universities, research institutions, experts, and volunteers.

H. Yokokawa
Graduate School of Agricultural and Life Sciences, The University of Tokyo, 1-1-1 Yayoi, Bunkyo-ku, Tokyo 113-8657, Japan

M. Mizoguchi (✉)
Graduate School of Agricultural and Life Sciences, The University of Tokyo, 1-1-1 Yayoi, Bunkyo-ku, Tokyo 113-8657, Japan

NPO "Resurrection of Fukushima", The University of Tokyo, Bunkyo-ku, Tokyo, Japan
e-mail: amizo@mail.ecc.u-tokyo.ac.jp

T.M. Nakanishi, K. Tanoi (eds.), *Agricultural Implications of the Fukushima Nuclear Accident*, DOI 10.1007/978-4-431-55828-6_16

This chapter examines the NPO "Resurrection of Fukushima" and reveals the structure of the collaboration system among residents, universities, research institutions, experts, and volunteers. In addition, the chapter focuses on the cooperation of the NPO and a group "Madei club," which consists of volunteers empathizing with the activity of the NPO and are working as staff of the Graduate School of Agricultural and Life Sciences at the University of Tokyo.

These cooperative frameworks are proposed as a realistic structure of collaboration among residents, universities, research institutions, experts, and volunteers.

16.2 NPO: Resurrection of Fukushima

The NPO was established in June 2011 by senior volunteers who live in Tokyo. The number of members was 261 on December 17, 2014. The members have various backgrounds such as retired office workers, active researchers, residents, and retired government officials. The core members of this NPO are elderly, around 70 years old (Fig. 16.1). The main activities are various field practice, which include measuring soil and air radioactivity; monitoring radiation, weather, and soil; developing decontamination methods; and trialing new industries for industrial reconstruction. The results from these practices are presented on the website (Resurrection of Fukushima).

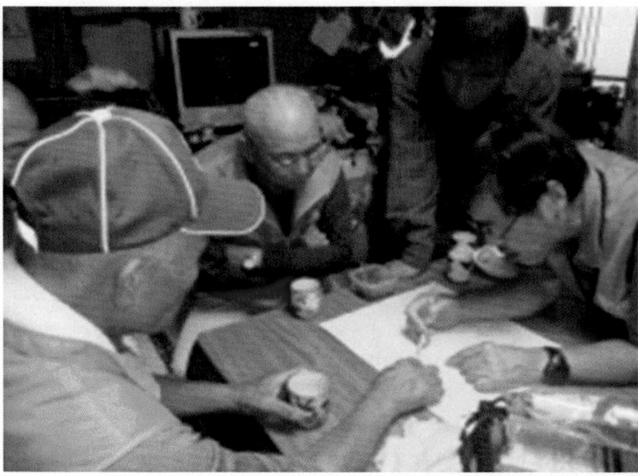

Fig. 16.1 Core members have a meeting

16.3 Collaboration Style of the Various Members

16.3.1 Collaboration Between Residents, Universities, and Research Institutions

Figure 16.2 shows the conceptual view of the collaboration in projects organized by "Resurrection of Fukushima." It has various professionals like researcher, medical worker, farmer, educator, journalists, and also the residents of the village. The NPO has connections with residents, universities, and research institutions, and it associates with them and plays a role to accelerate projects.

Case 1 Some members of this NPO are retired researchers. They form partnerships with research institutions easily by using their human network that they have built through their research. As a result, the NPO can ask the University of Tokyo and KEK (High Energy Accelerator Research Organization) to analyze the experimental samples containing radioactive materials (Arai et al. 2013).

Case 2 The members can use their expertise and technology to boost the ideas proposed by residents' ideas that are based on their local knowledge and lives. Specifically, they use experts at universities and research institutions to plan and conduct experiments that obtain data and evidence for realizing the residents' ideas.

Case 3 Some members are retired government officials who can deal with administrative procedures and issues, and thus enable projects to proceed smoothly.

These cases illustrate that the NPO supports the cooperation between residents, universities, and research institutions, and that they use the various backgrounds, human networks, experiences, and the expertise of various members.

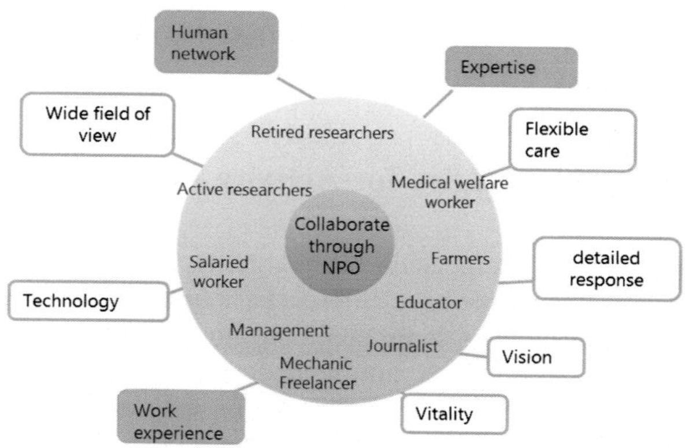

Fig. 16.2 Collaboration structure

Fig. 16.3 Poster of the "Madei club"

16.3.2 Cooperation with a Volunteer Group from the University of Tokyo

Samples collected in Iitate by NPO members are analyzed by researchers at the University of Tokyo. Each sample is packed into a small bottle for the analysis of radioactivity, which is time and labor consuming. To solve this problem, the group of volunteers in the "Madei club" assist with sampling and packing (Fig. 16.3). The volunteer group was established by staff at the university. The "Madei" is a local dialect word which means "heartful" in Iitate village. The volunteers sympathize with the activity of the NPO and want to do something for the village people. They volunteer during their lunch breaks and after their official work.

This is a significant case that illustrates the vast levels of cooperation within this project. This structure also enables volunteers who do not have a particular skill to support the NPO projects through the activities of the "Madei club".

16.4 The Motive for Activity

The motivation for action in this NPO is different for each person, particularly because the people have different backgrounds (Arai et al. 2013). Table 16.1 shows a classification of some common motives collected by interview. The motives include those which are related to helping people in Fukushima, such as "companionship" and "the sense of mission that our generation should take responsibility for the accident". Other motives that are important for participation include those related to enjoying the activities with "Resurrection of Fukushima" such as "fun to make new friends and meet new people".

Table 16.1 Classification of the motivations for participation

Motive	The voice of the member
Companionship	"I take part in the efforts because the victims and I are from the same hometown. I want to feel I can do something for them"
Feeling of pride in using the most advanced technology for decontamination	"I'm proud that our method of decontamination is the most advanced" "I feel that we are getting good results from decontamination"
Pleasure that their volunteer work matches with local people's demands	"I think there are many people that feel happy to work for other people's lives. Being helpful for other people's demands is a very important reason for participating in this volunteer organization"
A sense of mission that our generation should take responsibility for the accident	"I think it's difficult for young people to work for the reconstruction for their entire life, because they have many things they want to do. Old people have the responsibility for this problem and should deal with it"
Fun to make new friends and meet with new people	"It's fun to meet new people and make new friends. Something fun or interesting is an important factor for continuing our activity. Local people are also having fun participating"
To know the real situation	"To know what people really think in Fukushima, it's necessary to go there"

Table 16.2 Perspectives of the students who visited Iitate village and participate in "Resurrection of Fukushima"

Impression for	The voice of the visitor
"Resurrection of Fukushima"	"Resurrection of Fukushima is a necessary platform for everyone who wants to help the village"
	"Everyone is so lively and energetic (despite being old). I can't lose!"
Iitate village	"My initial image of the village was like a village without people. But there was such collaboration between people inside and outside"
	"Iitate looks like a ghost town. Everything is intact but deserted. It's a beautiful town, so it's a shame!"
	"I was shocked when I heard that some evacuated people were discriminated against because of unfounded fear of radiation"
What we should do	"The first step of support is to know and interact with local people" "What we should do is to visit and see the real situation" "It is important to tell the right insight and technology" "Our responsibility is to tell everyone the current situation of the village and activities of the NPO"
	"The studies of my major contributed with the reconstruction in the village. I work hard and I want to help them"

16.5 Perspectives of Visitors

Every year many people visit the village and participate in "Resurrection of Fukushima" to know the real situation in the contaminated area and to see the NPO activities (Arai et al. 2013; Osada 2013). In particular, many of the university students who cooperate with the NPO have visited. Table 16.2 presents the perspectives on the village and on the NPO from the students of University of Tokyo, Mie University, and Saga University.

16.6 Conclusion

While residents, universities, research institutions, and volunteers share the goal to reconstruct the disaster area, each member tends to act separately and the task of cooperating with each other is difficult. In these cases, however, the various members can cooperate smoothly through the NPO "Resurrection of Fukushima". The most important key for collaboration is the connection of people. The group of people who have various backgrounds, experiences, human networks, and motivations is the most suitable mediator to connect separate members.

The disaster area looks like a "ghost town", because the terrible disaster deprived the town of many important things. However, a new style of cooperation in research and in practice with researchers, volunteers, and local residents has arisen from such a situation. The style will become a seed for new agricultural sciences in the future.

Acknowledgement We sincerely thank Mr. Muneo Kanno, a farmer at iitate Village, and Mr. Yoichi Tao, the representative of the NPO "Resurrection of Fukushima", for their kind collaboration in our activities. Some of the activities and works are available at the website: http://www.iai.ga.a.u-tokyo.ac.jp/mizo/edrp/fukushima/Fukushima_articles.html

Supplemental Materials

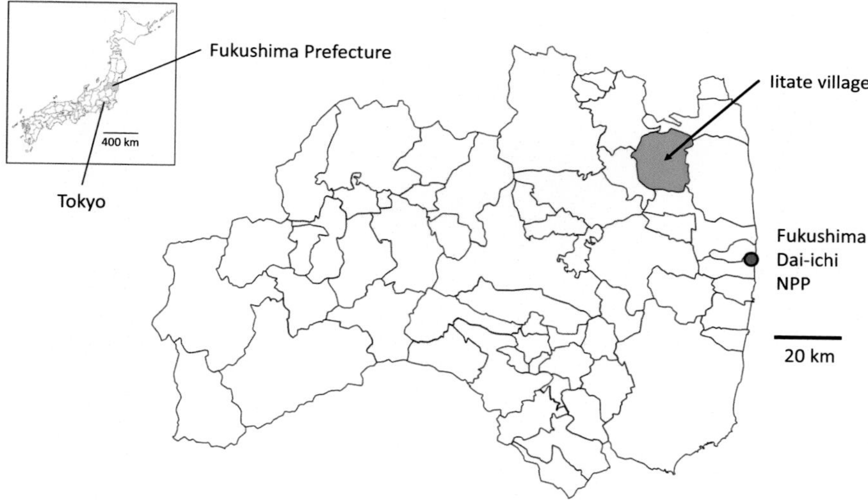

Location of Iitate village

Photos of Activities in Iitate Village

Soil sampling under snow by collaboration of different generations (Feb. 17, 2013)

Soil cutting work by "Madei club" members who are working in University of Tokyo (Feb. 24, 2015)

Decontamination by forced drainage of puddling muddy water by collaboration of a local farmer and NPO members. (May 18, 2013)

Handshake by local farmers, NPO and university after harvesting rice on decontaminated paddy field (Oct. 6, 2013)

The NPO representative, Mr. Tao, is reporting the result of survey in Residents briefing (Apr. 12, 2015)

The NPO representative, Mr. Tao, is explaining its activity in a paddy field after harvesting in Iitate for young generation (Oct. 27, 2013)

References

Arai K, Ota T, Soni P, Yokokawa H, Watanabe L (2013) Research as a volunteer – a case study of the resurrection of Fukushima. Group 6, Department of Global Agricultural Sciences. www.iai.ga.a.u-tokyo.ac.jp/mizo/lecture/noukoku-1/group-work/2012/G6_e.pdf

Osada Y (2013) Excursion tour in Fukushima held after the 55th symposium. J Jpn Soc Soil Phys 125:55–57, https://js-soilphysics.com/downloads/pdf/125055.pdf

Resurrection of Fukushima (Fukushima Saisei no Kai). http://www.fukushima-saisei.jp/en/en-purpose/Collection of Mizo's works on Fukushima, http://www.iai.ga.a.u-tokyo.ac.jp/mizo/edrp/fukushima/Fukushima_articles.html

Chapter 17
Impacts of the Nuclear Power Plant Accident and the Start of Trial Operations in Fukushima Fisheries

Nobuyuki Yagi

Abstract The large-scale release of radioactive substances from the Fukushima Daiichi Nuclear Power Plant operated by the Tokyo Electric Power Company in March 2011 caused significant damage to local fisheries. The Fukushima Prefectural Federation of Fisheries Cooperative Associations immediately suspended all commercial fishing activities within Fukushima. The national government issued instructions prohibiting the sale of certain marine products caught in the waters off Fukushima Prefecture due to food safety concerns. The prohibition is gradually being lifted; in June 2012, the Fisheries Cooperative Associations resumed commercial fishing of three species (two octopus species and one shellfish species) as trial operations. Subsequently, the list has expanded, and as of January 2015, 58 species have been approved for trial operations. The scale of operations is far smaller than before the tsunami and nuclear accident. A full recovery of Fukushima fisheries cannot be realized until the government lifts prohibitions on the sale of all remaining marine species, and the timing of such a decision remains unclear.

Keywords Cesium • Fukushima • Radioactive substances • Trial fishery

17.1 Introduction

The tsunami on March 11, 2011 damaged around 29,000 fishing boats and 319 fishing ports in Japan (Fisheries Agency of Japan 2014a). Each of these figures accounts for approximately 10 % of the respective national totals. Since October 31, 2014, approximately 17,713 of these boats and 302 of the ports have again become operational (Fisheries Agency of Japan 2014a). Despite the unprecedented scale of the disaster, the rehabilitation of the fisheries in the tsunami-damaged areas, excluding Fukushima Prefecture, has been relatively expeditious in terms of fishing capacity as measured by the number of boats and ports (Yagi 2014).

N. Yagi (✉)
Graduate School of Agricultural and Life Sciences, The University of Tokyo, 1-1-1 Yayoi, Bunkyo, Tokyo 113-8657, Japan
e-mail: yagi@fs.a.u-tokyo.ac.jp

T.M. Nakanishi, K. Tanoi (eds.), *Agricultural Implications of the Fukushima Nuclear Accident*, DOI 10.1007/978-4-431-55828-6_17

In Fukushima Prefecture, significant damage has been caused by the large-scale release of radioactive substances from the Fukushima Daiichi Nuclear Power Plant operated by the Tokyo Electric Power Company (TEPCO). The recovery of fishing capacity has been slower in Fukushima than in the other two tsunami-damaged prefectures, namely Iwate and Miyagi. A total of 873 fishing vessels were damaged by the tsunami in Fukushima Prefecture (Ministry of Agriculture, Forestry and Fisheries, Japan 2012). As of October 31, 2014, 327 fishing vessels and 8 fishing ports in Fukushima Prefecture have been repaired or reconstructed (Fisheries Agency of Japan 2014a). Nonetheless, as of January 2015, most fisheries in Fukushima have not been able to resume their commercial operations due to nuclear damage.

This chapter of the book reviews the state of fisheries in Fukushima since the nuclear power plant accident. It also provides precise records regarding the start of trial operations in June 2012. Finally, the discussion section is provided that considers future options for fisheries in Fukushima.

17.2 Commercial Fisheries in Fukushima Before and After the Disaster

Within the waters of Fukushima,[1] roughly 10 different fisheries were being operated until the nuclear accident in 2011. These fisheries included bottom trawling, mid-water trawling, purse seine, gill net, saury dip net, trap net, long line, line and pole (jigging), pot fishery, and hand gathering. The national and prefectural governments set a licensing system and various regulations, such as total allowable catches (TACs) for several species, restrictions on gear use, and restricted areas of operations for each fishery. While purse seine and saury dip net fisheries can be operated both inside and outside the waters of Fukushima, the other eight fisheries were in principle only allowed inside the waters of Fukushima.

The average landing volume of fisheries in Fukushima before the disaster was approximately 50 thousand metric tons per year (Fukushima Prefecture 2011, 2012). Of these, bottom trawling, mid-water trawling, and purse seine fisheries were the three major fisheries. Table 17.1 provides an overview of the areas of operations, the main target species, and the average annual landing volumes before the nuclear accident.

Figure 17.1 provides time series data on annual landings. On average, approximately 40–50 thousand metric tons of annual landings were recorded by Fukushima fisheries before the disaster, but this decreased to less than 5 thousand metric tons after the nuclear accident (Fukushima Prefecture 2011, 2012, 2013). Landings from

[1] The term "waters of Fukushima" used in this chapter refers to ocean areas bounded by the coastline of Fukushima Prefecture, the outer limit of Japan's Exclusive Economic Zone, and the boundaries with Miyagi and Ibaraki prefectures (see Fig. 17.2).

Table 17.1 Fisheries in Fukushima before the nuclear accident

Type of fishery	Main target species	Area of operation	Annual landing (in metric tons)
Bottom trawling	Pacific cod, flatfish, octopus	Offshore	10,588
Mid-water trawling	Sand lance larvae (whitebait)	Coastal	10,156
Purse seine	Skipjack tuna, mackerel	Offshore/ outside	18,253
Gill net	Spanish mackerel, yellowtail	Coastal/ offshore	2496
Saury dip net	Pacific saury	Offshore/ outside	5803
Trap net	Chum salmon, sea bass	Coastal	276
Long-line	Pacific cod	Coastal	207
Line and pole, jigging	Skipjack tuna, bigeye tuna	Coastal/ offshore	1111
Pot fishery	Octopus, shellfish	Coastal/ offshore	1098
Hand gathering	Shellfish, seaweed	Shoreline	84

Source: Compiled by the author based on information available from the Fukushima Prefectural Government

"Annual landing" refers to the average annual landing reported from 2001 to 2010 at ports in Fukushima Prefecture. The figures include landings by Japanese fishing boats registered in both Fukushima and other prefectures in Japan. They do not include landings by Fukushima vessels at ports outside of Fukushima Prefecture. "Offshore/outside" refers to offshore areas of waters in Fukushima and other prefectures in Japan. "Offshore/coastal" refers to both offshore and coastal areas in Fukushima Prefecture

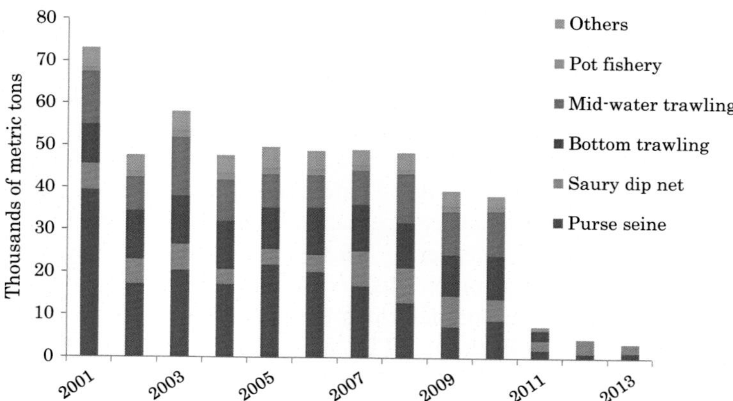

Fig. 17.1 Commercial landings from fisheries in Fukushima. These figures include landings by Japanese fishing boats registered in both Fukushima and other prefectures in Japan. They do not include landings by Fukushima vessels at ports outside of Fukushima Prefecture. Landings from "trial fishing" after June 2012 are not included (*Source*: Fukushima Prefecture 2011, 2012, 2013)

2012 to 2013 were harvested with purse seines and saury dip net fisheries outside the waters of Fukushima and then landed in the ports of Fukushima Prefecture. The landings in 2011 include ordinary commercial harvests from January to March 11, 2011.

17.3 Initiation of Trial Operations in Fukushima Fisheries

As a result of the large-scale release of radioactive substances from the Fukushima Daiichi Nuclear Power Plant following the earthquake and tsunami of March 11, 2011, the Fukushima Prefectural Federation of Fisheries Cooperative Associations (hereafter referred to as the Fukushima FCA) decided on March 15, 2011 to voluntarily stop fishing operations in the waters of Fukushima (Yagi 2014). Some fishing activities in the prefectures neighboring Fukushima (namely Miyagi and Ibaraki) were also suspended after the TEPCO accident, but most of these were subsequently reopened within 2 years (Ibaraki Prefecture 2014; Miyagi Prefecture 2014).

Neither the national nor the prefectural governments revoked fishing licenses in Fukushima. However, the national government did provide legally binding sales prohibitions on certain marine products caught in the waters off Fukushima Prefecture based on the food safety requirements. The Japanese government set the allowable level of radioactive cesium for all fisheries products as 500 Bq/kg until it was reduced to 100 Bq/kg on April 1, 2012 (Fisheries Agency of Japan 2014b). In addition, species-specific prohibitions on sales and marketing for specific agricultural and fishery products, originating from certain areas were introduced by the government regardless of the actual measured levels of radioactive substances.

Many of the fishers in Fukushima have been receiving a certain level of compensation from TEPCO since the nuclear accident (Yagi 2013). According to the law, the nuclear power plant operator TEPCO should compensate fishers for stopping their operations in Fukushima (Fisheries Agency of Japan 2012). Although TEPCO is paying compensation, many fishers have complaints. Some fishers are unable to submit evidence that could be used to calculate the amount of actual damage, because sales slips dated before the accident were lost during the tsunami, resulting in inappropriate compensation to such fishers. Processing and/or distribution companies are considering leaving Fukushima, as they cannot get adequate compensation. Furthermore, fishers can neither estimate how long they need to cease their operations nor make long-term plans for the future (Yagi 2013).

Within this context, the Fukushima FCA launched the "Fukushima Prefectural Fisheries Reconstruction Committee" in 2012 with the aim of reconstructing the fisheries industry and restarting fisheries operations. The author is a member of this committee and took note as a participant that the committee discussed issues including the following (Yagi 2014).

- From January 1, 2012 to June 4, 2012, Fukushima Prefecture analyzed 2118 samples of marine organisms collected from the waters of Fukushima. The survey showed that the level of radioactive substances in fish is site-specific. The levels were higher in samples collected near the nuclear plant and became lower in samples collected in deeper ocean areas. The levels of radioactivity are also species-specific and invertebrate species contained less cesium in 2011 and levels dropped to zero in 2012.
- On the other hand, as of June 2012, more than 100 Bq/kg of radioactive cesium was still being recorded in Japanese sea bass and other fish species living along the coast.
- An independent biological study has showed that fish species have their own function for excreting cesium from their bodies (Furukawa et al. 2012).
- These findings are consistent with the International Atomic Energy Agency (IAEA) report that shows marine invertebrate animals have low concentration factors compared with fish (figures calculated by dividing the level of cesium contained in the body by the level of cesium contained in the water) (IAEA 2004).

In June 2012, after taking into account the above information considered by the Fukushima Prefectural Fisheries Reconstruction Committee, the Fukushima FCA decided to resume fishing activities, named as "trial operations," for three species (two octopus species and one shellfish species) living at depths of more than 150 m in ocean areas approximately 60–90 km from the damaged nuclear power plant. Because the government did not revoke fishing licenses for fishers in Fukushima after the nuclear power plant accident, the fishers were left to decide for themselves whether to resume commercial fisheries by targeting species that met government food safety standards.

The trial operation has several limitations, and therefore it is not regarded as a full resumption of commercial fisheries. The limitations include: (1) days of operation (usually fewer than 5 days a month); (2) landing ports (only two ports have been designated: one in Soma and the other in Iwaki); (3) the amount of landed fish (usually less than 10 tons a day); and (4) the number of vessels involved in fishing operations (Yagi 2014). Operation rules are set in order to maintain a high frequency of monitoring for radioactive substances and to ensure traceability following landing of marine products.

Figure 17.2 shows the areas for trial operations. Initially the trial operations for bottom trawling were only allowed in Area 1. This area was expanded to include Area 2 in October 2012, Area 3 in February 2013, Area 4 in May 2013, Area 5 in August 2013, and Area 6 in December 2013. The pot fishery was limited to Areas 1, 2, 3, and 4 in 2013. Trial operations for mid-water trawling started in March 2013 and are allowed in Area A. In February 2014, such operations were added to Area B.

The Fukushima FCA for a certain number of samples conducts monitoring of radioactive cesium at landing sites. These samples are randomly selected, but they do not cover every fisheries product, and because the testing method involves

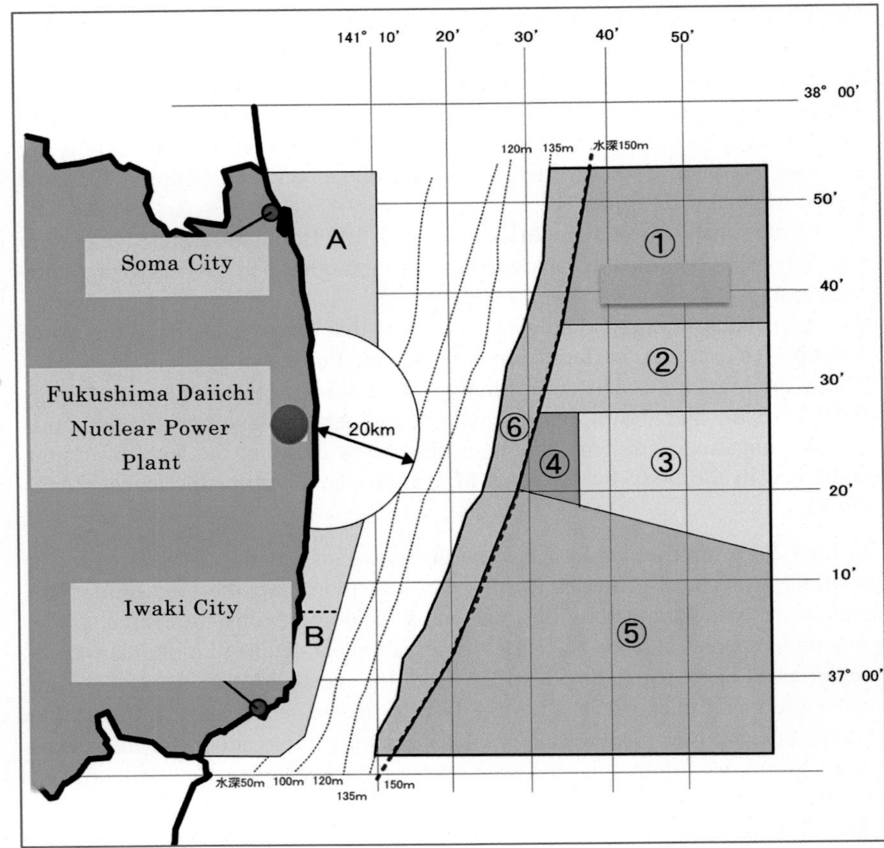

Fig. 17.2 Areas for trial operations (*Source*: Fukushima FCA http://www.fsgyoren.jf-net.ne.jp/siso/buhin/kaiiki20140827.pdf)

grinding up the fish, the tested samples are not marketable afterwards. When any product exceeds the Fukushima FCA's voluntary standard level,[2] which is 50 Bq/kg of radioactive substances, all products are immediately recalled (Yagi 2014).

Landed octopus and shellfish have been sold with labels indicating Fukushima as their point of origin (Yagi 2014). Most of these products were sold at local supermarkets in Fukushima, and sold out very quickly, most likely due to the small quantity of available items and a number of consumers wishing to help their local fishers by purchasing their products (Yagi 2014).

Figure 17.3 indicates the scale of the trial operations. As of January 2015, the trial operations are only being allowed for bottom trawling, mid-water trawling, and

[2] Although the threshold set by the government is 100 Bq/kg, the Fukushima FCA decided to set a more conservative limit of 50 Bq/kg as its voluntary standard.

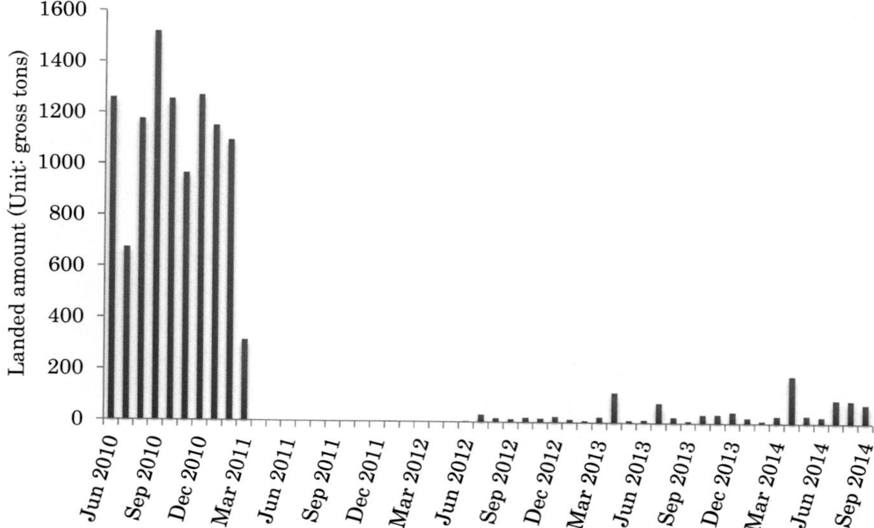

Fig. 17.3 Monthly landings of commercial fisheries before the nuclear accident in March 2011 and since trial operations started in June 2012. Data were compiled by the author using information available from Fukushima Prefecture

pot fishery. The monthly average landing of the three fisheries used to be around 1200 gross tons before the disaster in March 2011. The landings under the trial operations, however, are around 40 gross tons per month, or approximately 1/30th of those recorded before the disaster.

The committee also discussed how some consumers are likely to object to the limited resumption of fishing activities due to concerns that fishing operations should not be allowed at all when there is a risk of consumers ingesting radioactive substances contained in fish products (Yagi 2014). Some consumers may also object to only checking radioactive substances in some samples and not in the full catch. The committee argued that "Each individual has a right to select what they want. Rather than trying to persuade consumers that Fukushima products on the market are safe, it is better to clearly label the products as originating from Fukushima and to allow consumers to make their own decisions about consumption behavior" (Yagi 2014).

The number of species allowed under the trial fishery operations gradually expanded in 2012: seven species were added in August, and another three species were added in November. An additional 18 species were added to the list over the course of 2013. Thus, a total of 31 species were subject to trial fishery operations as of February 2014 (Fukushima FCA 2014). The number subsequently increased, and 58 species are now allowed under the trial fishery operations as of January 2015.

17.4 State of Radioactive Substances in Marine Species

Since the nuclear power plant accident in March 2011, various governmental institutes have collected and analyzed samples of marine organisms and released information on levels of radioactive substances. One of the most extensive datasets for the testing results is available through the website of the Fisheries Agency of the Government of Japan (http://www.jfa.maff.go.jp/j/housyanou/kekka.html). The data includes information on each analyzed sample with respect to its species name, sampling area, its level (Bq/kg) of radioactive cesium 134 and 137, date of sampling, and the name of the institution that analyzed the sample. Data are available for 27,283 samples collected inside Fukushima Prefecture and another 39,228 from outside of Fukushima Prefecture. Of the samples collected outside Fukushima, approximately 6.5 % of the samples of marine species contained radioactive cesium above 100 Bq/kg between March and June 2011. From July 2011 to March 2012, approximately 5 % of the samples contained radioactive cesium above 100 Bq/kg. The proportion of samples that contained radioactive cesium above 100 Bq/kg gradually decreased during 2012. Since 2013, less than 1 % of samples collected outside the waters of Fukushima contained radioactive cesium above 100 Bq/kg.

The status of samples collected from within the waters of Fukushima is somewhat different (see Fig. 17.4). Fukushima Prefecture has regularly been collecting samples of marine species from all areas across the waters in Fukushima since March 2011. In April 2011, 13 samples were collected, and it was found that 12 of them (92 %) contained radioactive cesium over the level of 100 Bq/kg (Fukushima Prefecture website: http://www.pref.fukushima.lg.jp/sec/36035e/suisanka-monita-top.html). Thereafter, 32 samples out of 47 (68 %) in May, 72 samples out of 145 (50 %) in June, 95 samples out of 201 (47 %) in July, and 103 samples out of 238 (43 %) samples in August contained radioactive cesium in excess of 100 Bq/kg. During this early period after the disaster, the monthly sample size was less than 300, but this has subsequently increased to around 700 samples per month (see Fig. 17.4). From September 2011 to April 2012, more than 20 % of the sampled marine species contained radioactive cesium over 100 Bq/kg. For the first half of 2013, 3–8 % of samples contained cesium over 100 Bq/kg, and after July 2013, the proportion dropped to 0–2 %.

Detailed outcomes of the tests above are also available from Wada et al. (2013). Several other analyses have also been published by independent researchers from outside the government. Kikkawa et al. (2014), for instance, used published data for 8683 samples of aquatic animals and plants obtained off the coast of Fukushima from 2011 to 2013 and evaluated the level of radioactive cesium (^{134}Cs and ^{137}Cs) for 95 species (a total of 97 fishery items including two species that are marketed separately at the adult and immature stages). A cluster analysis based on parameters of annual average and standard deviation of radioactive cesium concentration levels indicated that the 97 items can be categorized into four groups. The first group had lower concentrations and lower variability across the first and second years

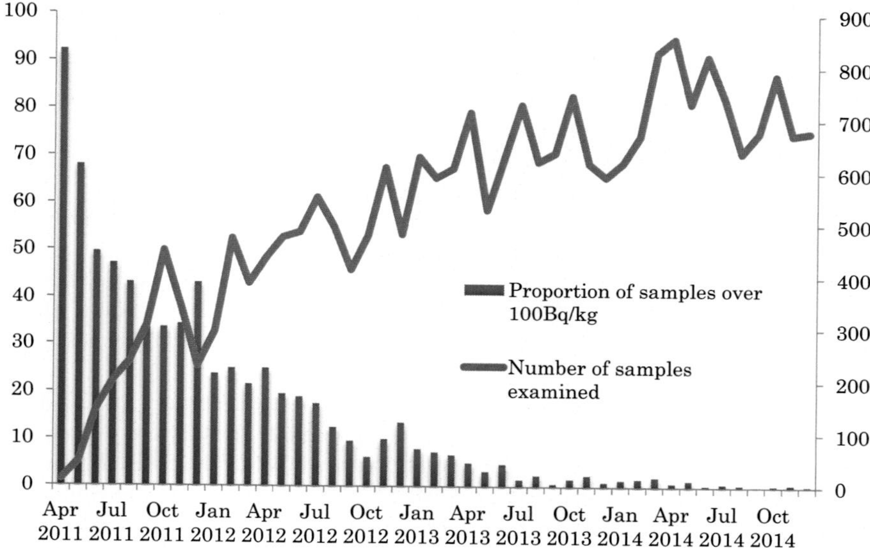

Fig. 17.4 Levels of radioactive cesium in marine species sampled in the waters of Fukushima between April 2011 and December 2014. The *left axis* indicates the percentage of samples that contained over 100 Bq/kg of radioactive cesium. The *right axis* shows the number of samples examined per month. The figure is compiled by the author using data available from the Fukushima Prefectural Government http://www.pref.fukushima.lg.jp/sec/36035e/suisanka-monita-top.html

(60 items); the second group showed a decline in concentration levels in the second year, but remained high (21 items); the third group had extremely high initial concentrations, but levels became almost undetectable in the second year (1 item); and the fourth group exhibited high contamination levels across the 2 years (15 items) (Kikkawa et al. 2014). The researchers argue that almost all of the items in the first and third groups satisfied the government's food safety standards in the second year. On the other hand, products in the second and the fourth groups do not satisfy these standards, and must be closely monitored (Kikkawa et al. 2014).

The names of species that are subject to sales prohibitions have been announced by the Government of Japan. The sales prohibition is legally binding and the government periodically revises the list based on the best available information. Since December 2013, the prohibition has been in effect for 40 marine species living in the waters of Fukushima (Fukushima Prefecture 2014), and since January 2015, it has been reduced to 35 species (Fukushima Prefecture 2015).

On the other hand, the target species for the trial operations are decided by the Fukushima FCA in consultation with the Fukushima Prefectural Fisheries Reconstruction Committee, and they discuss the most recent testing results for radioactive substances in all marine species. As of January 2015, 58 species are allowed for trial operations. In sum, approximately 200–300 marine fish and invertebrate species occur in the waters of Fukushima, and 58 of these are targeted by the trial operations, while 35 other species are prohibited for sale due to government regulations. The rest are still subject to monitoring for radioactive substances by research institutions, but are not treated as species for trial operations.

As mentioned earlier in this chapter, all products are immediately recalled if any product exceeds 50 Bq/kg of radioactive cesium. On February 27, 2014, such an incident happened for the first time. A sample taken from one particular species of ocean perch (*Helicolenus hilgendorfi*: Japanese name *yumekasago*) was measured as having 112 Bq/kg of radioactive cesium. The sample was measured using a regular checking process conducted by the Fukushima FCA at the landing site in Iwaki. The Fukushima FCA immediately decided to stop distributing *yumekasago* from Iwaki.

The same species was also landed on the previous day (February 26, 2014) at Soma, which is the other port designated as a landing site for the trial operations. Although *yumekasago* samples inspected at Soma were below the 50 Bq/kg threshold, the Fukushima FCA decided to recall all *yumekasago* landed at Soma the previous day from the middlemen and retail stores. According to the records of the Fukushima FCA, 33.5 kg of *yumekasago* were landed on February 26, 2014. Some had already been sold to consumers, but retail stores posted a message to consumers that they would recall the product and provide the corresponding refund.

All *yumekasago* from the waters of Fukushima disappeared from the market from 27 February until September 1, 2014, when the Fukushima FCA decided to restart trial operations and distribution for this species based on the consensus decision of the Fukushima Prefectural Fisheries Reconstruction Committee, which reviewed the outcome of recent testing of radioactive cesium contaminations in samples of *yumekasago*.

17.5 Future Perspectives for Fukushima Fisheries

The Fukushima FCA started its trial operations in June 2012 to target species occurring in offshore areas. It subsequently included species from the mid-surface layer of the coastal areas outside the 20 km radius from the nuclear power plant. The landing amount of the trial operations is around 1/30th of that harvested by commercial fisheries before the disaster. The full recovery of commercial fisheries in Fukushima has yet to be achieved. As of January 2015, the sale and marketing of 35 marine species are prohibited (Fukushima Prefecture 2015). These species typically live on the sea bottom of very near-shore areas. Full recovery of Fukushima fisheries cannot be realized until the government lifts the sales prohibition on all remaining marine species, but the timing of such a decision remains unclear.

At the same time, there are several issues that need to be addressed. These include issues related to weak consumer confidence following the nuclear power plant accident. Numerous incidents of weak consumer confidence in fisheries products from the tsunami-affected areas have been reported in newspapers and other media in Japan. Increased communication between producers and consumers assist consumer confidence. In order to increase communication between producers and consumers, it could be effective to shorten the distribution channel. Domestic fish distribution in Japan is composed of multiple layers of traders with two stages of wholesale markets: at the landing site and the consumption site. The first landing site handles the harvested fish and includes middlemen and distributors, while the consumption site wholesale market (e.g., Tsukiji Market in Tokyo) is located in cities and includes wholesalers and brokers. Retailers are then added to the value chain. Information about the production site is hard to communicate to consumers across this long chain. Improved communication and the establishment of accurate traceability mechanisms in the value chains can play a role in reducing consumer concerns (Yagi 2013).

Reforming the value chains for fishery products can help to reverse the current weak position of producers in the value chain compared with retailers (Sakai et al. 2012; Nakajima et al. 2011, 2014), and it can provide a solution to the economic weakness of Japan's fish harvesting sectors as a whole that pre-dates the March 2011 disaster. Even before the tsunami, the Ministry of Agriculture, Forestry and Fisheries pointed out that, while producers receive 24.7 % of the retail price of the fish, retailers receive 38.5 % and middlemen/distributers receive the remaining share (36.8 %) (Fisheries Agency of Japan 2009). It is extremely difficult to raise the price of fishery products in retail markets in the midst of weak consumer confidence. To enlarge the share of producers in the value chain, one possible option could take the form of new e-commerce business mechanisms to sell products more efficiently to urban consumers; any other forms of direct marketing to consumers would also be useful.

References

Fisheries Agency of Japan (2009) The fisheries white paper in 2009 issued by the Fisheries Agency of Japan. http://www.jfa.maff.go.jp/j/kikaku/wpaper/h21/gaiyou.html. Accessed Feb 2014 (in Japanese)

Fisheries Agency of Japan (2012) Q & A on the power-plant accident. Homepage of fisheries agency. http://www.jfa.maff.go.jp/j/sigen/gensiryoku/index.html. Accessed Aug 2012 (in Japanese)

Fisheries Agency of Japan (2014a) The great East Japan earthquake's impact on fisheries and future measures. http://www.jfs.maff.go.jp. Accessed Dec 2014 (in Japanese)

Fisheries Agency of Japan (2014b) Questions and answers on fisheries. http://www.jfa.maff.go.jp/j/kakou/Q_A/index.html. Accessed Feb 2014 (in Japanese)

Fukushima Prefecture (2011) Marine fishery production statistics of 2010. http://www.pref.fukushima.lg.jp/sec/36035e/suisanka-toukei-top.html. Accessed Jan 2015 (in Japanese)

Fukushima Prefecture (2012) Marine fishery production statistics of 2011. http://www.pref.fukushima.lg.jp/sec/36035e/suisanka-toukei-top.html. Accessed Jan 2015 (in Japanese)

Fukushima Prefecture (2013) Marine fishery production statistics of 2012. http://www.pref.fukushima.lg.jp/sec/36035e/suisanka-toukei-top.html. Accessed Jan 2015 (in Japanese)

Fukushima FCA (2014) A list of marine species permitted for the limited experimental operations in Fukushima fisheries. http://www.jf-net.ne.jp/fsgyoren/siso/buhin/taisyousyu.pdf. Accessed Feb 2014 (in Japanese)

Fukushima Prefecture (2014) Homepage of Fisheries Division, Fukushima prefecture. http://wwwcms.pref.fukushima.jp/pcp_portal/PortalServlet. Accessed Feb 2014 (in Japanese)

Fukushima Prefecture (2015) Discussion material for "Fukushima Prefectural Fisheries Reconstruction Committee." Materials prepared by Fukushima prefecture and distributed to the press and committee members on 22 Jan 2015

Furukawa F, Watanabe S, Kaneko T (2012) Excretion of cesium and rubidium via the branchial potassium-transporting pathway in Mozambique tilapia. Fish Sci 78:197–602

Ibaraki Prefecture (2014) Restricted items for marketing and distributions as of February 21, 2014. http://www.pref.ibaraki.jp/nourin/gyosei/housyanou_jyouhou.html#4. Accessed Feb 2014 (in Japanese)

International Atomic Energy Agency (2004) Sediment distribution coefficients and concentration factors for biota in the marine environment. Technical report 422, Vienna

Kikkawa T, Yagi N, Kurokura H (2014) The state of concentration of radioactive cesium in marine organisms collected from the Fukushima coastal area: a species by species evaluation. Bull Jpn Soc Fish Sci 80:27–33 (in Japanese)

Ministry of Agriculture, Forestry and Fisheries, Japan (2012) Basic statistical data on damages in agriculture, forestry and fisheries caused by the Great East Japan Earthquake. http://www.maff.go.jp/j/tokei/joho/zusetu/pdf/00_2406all.pdf. Accessed Jan 2015 (in Japanese)

Miyagi Prefecture (2014) Restricted items for marketing and distributions as of February 18, 2014. http://www.pref.miyagi.jp/uploaded/attachment/245220.pdf. Accessed Feb 2014 (in Japanese)

Nakajima T, Matsui T, Sakai Y, Yagi N (2011) An approach to capture changes in asymmetric price transmission: rolling window TAR estimations using the blue fin tuna data. J Int Fish 10:1–18

Nakajima T, Matsui T, Sakai Y, Yagi N (2014) Structural changes and imperfect competition in the supply chain of Japanese fisheries product markets. Fish Sci 80:1337–1345

Sakai Y, Nakajima T, Matsui T, Yagi N (2012) Asymmetric price transmission in the Japanese seafood market. Bull Jpn Soc Fish Sci 78:468–478 (in Japanese)

Wada T, Nemoto Y, Shimamura S, Fujita T, Mizuno T, Sohtome T, Kamiyama K, Morita T, Igarashi S (2013) Effects of the nuclear disaster on marine products in Fukushima. J Environ Radioact 124:246–254

Yagi N (2013) Challenges faced with fishing industries in Fukushima and their long-term solutions. Bull Jpn Soc Fish Sci 79:88–90 (In Japanese)

Yagi N (2014) The state of fishing industry in Fukushima after the nuclear power plant accident. Glob Environ Res 18:65–72

Chapter 18
Consumer Evaluation of Foods from the Disaster Affected Area: Change in 3 Years

Hiromi Hosono, Yuko Kumagai, Mami Iwabuchi, and Tsutomu Sekizaki

Abstract Since the Great East-Japan Earthquake and the following nuclear power plant accident, consumer anxiety on food and environmental contamination of radioactive substances have spread widely. This chapter examines the change in consumer attitude towards foods, focusing on beef, produced in disaster affected area based on the 6 times of web-based survey from 2011 October to 2014 March. The results showed that the risk of radioactive substances through beef consumption are not regarded as high as microbial hazards. And trust on radiation risk management implemented by government as well as food business were recovering. However, the ratio of those who stated zero WTP for foods produced in disaster affected area were rather increasing or remain constant since 2012. The results of choice experiment indicated providing information of radiation risk and risk management is effective to recover WTP for beef produced in Fukushima while knowledge level remained relatively low. We believe that continuous and accessible communication with consumers would contribute to the recovery of devastated area.

Keywords Consumer behavior • Risk perception • Trust • Knowledge • Information

18.1 Introduction

More than 3 years have passed since the Great East Japan Earthquake and TEPCO's first Fukushima Daiichi Nuclear Power Plant explosion that led to catastrophic damage to East Japan. The disaster caused huge damage to the social life of the Japanese, especially in eastern Japan. Immediately after the accident, consumers were concerned about radioactive substances polluting the environment and food. A

Hiromi Hosono was deceased at the time of publication.

H. Hosono • Y. Kumagai • M. Iwabuchi • T. Sekizaki
Graduate School of Agricultural and Life Sciences, The University of Tokyo, 1-1-1 Yayoi, Bunkyo-ku, Tokyo 113-8657, Japan

T.M. Nakanishi, K. Tanoi (eds.), *Agricultural Implications of the Fukushima Nuclear Accident*, DOI 10.1007/978-4-431-55828-6_18

considerable number of people were forced to reside in areas some distance from the affected area, and people started to pay more attention to the origin of their food to avert health risks.

The Japanese Ministry of Health, Labour and Welfare (MHLW) adopted a provisional regulation for the level of radioactive substances in foods on March 17, 2011. In April 2011, this regulation was revised upward and a maximum permissible dose of radioactive substances was set for each food category. The acceptable value for radioactive cesium in general food products was designated to be 100 Bq/kg or less. Along with setting standards, intensive inspection of atmosphere, soil, and agricultural products was initiated to assess the contamination situation and to plan and implement decontamination activities as well as to restrain the shipping of foods contaminated with excess levels of radioactive substances. With these compositive activities, most foods that were inspected were below the regulation level of radiocesium contamination, except for wild animal or vegetables and marine products.

However, consumer anxiety about foods produced in the disaster-affected area is not expected to disappear in the short term. According to the annual consumer research implemented by the Food Safety Commission in August 2013, 29.5 and 38.0 % of responders felt very uneasy and uneasy, respectively, about the risk of radioactive substances in foods. The ratio of feeling uneasy ranked the second highest after microbial contamination (20.7 and 60.1 %) among 12 food-related hazards, but when the focus was on those who felt very uneasy, radioactive substances ranked the highest. Compared with the survey in 2011 and 2012, the ratio of those who felt very uneasy and uneasy about food contamination with radioactive substances was decreasing; 88.5, 80.3, and 74.2 % in August 2011, March 2012, and September 2012, respectively, although considerable numbers of people were still anxious about the risk.

Other consumer research also indicated that there was broad consumer anxiety about radiation risk (Kito 2012; Hangui 2013; Kurihara et al. 2013). Ujiie (2012, 2013) conducted research every 3 months about consumer acceptance of foods produced in Fukushima and Ibaraki prefectures between March 2011 and February 2013. The price differences in produce between these two sites and another area have been regarded as a willingness to accept (WTA) the situation. In Ujiie's analysis, WTA's were separated into "health risk estimation due to radioactive contamination" and "effect of production at the site." As a result, WTA increased in August 2011, which may have been due to a high dose of radiocesium detected in beef. Yoshida (2013) conducted two consumer surveys (January and December 2012) and warned of the possibility of a regime shift regarding foods from the disaster-affected area; the shift can occur in people experiencing catastrophic shock beyond resilience. Research by other consumers also noted the high anxiety, high-risk perception, and difficulty of recovering consumer confidence.

In this article, we describe the change in Japanese consumer attitudes, knowledge, risk perception, and food-purchasing behavior based on a series of web-based consumer surveys implemented from 2011 to 2014.

18.2 Research Outline

We conducted consecutive web-based consumer surveys from October to November 2011 (N = 4363), March 2012 (N = 5028), January 2013 (N = 6357), and February 2014 (N = 9678) to investigate consumer attitudes and knowledge on radioactive substances, risk perception, and risk management measures taken in Japan with a focus on beef. The first two surveys were monitored by Nikkei Research Inc. and the others were monitored by Nippon Research Center, Ltd. The respondents included both males and females recruited from all prefectures aged in their 20s–60s. After the 3rd and the 5th (last) surveys shown above, a web-based donating experiment was offered to the 1881 (4th) and 1822 (6th) participants, respectively, selected from the 3rd and 5th survey responses. In this experiment, the participants received 100–10,000 JPY as the result of the Ultimatum Game. Using this money, they were then asked whether to donate or not. Ten charity organizations were provided as options, including enhancement of food radiocesium monitoring, compensation for farmers, subsidization for recovery from the tsunami-affected area, research on radiation risk, and development of new energy resources.

The 1st to 3rd and 5th surveys covered the following subjects: (1) perceived risk level of eight (or seven) beef-related hazards; (2) knowledge about food safety risks and risk management focusing on radiocesium; (3) attitude toward food safety and radiation risk management; (4) intention to support rehabilitation and reconstruction from the disaster; (5) intention to purchase food produced in East Japan and Fukushima prefecture; and (6) demographic characteristics such as age, gender, residential area, and household members. We also implemented the beef choice experiment on the 3rd and 5th survey and risk management measure choice experiment on the 4th and 6th survey. Before the beef choice experiment, some respondents were provided with information about radiation risk and the risk management measures conducted after the disaster. The information included risk of radioactive cesium from foods, standards and risk reduction measures taken in Japan, and the results of food radiation inspections. The demographic characteristics of the respondents are shown in Table 18.1. In the following section, we mainly focus on the 3rd and 5th surveys and the subsequent donating experiment.

18.3 Results

18.3.1 Risk Perception

Eight hazards were itemized regarding assumed degrees of health risk originating in beef: enterohemorrhagic *Escherichia coli*, *Salmonella* sp., *Campylobacter* spp., antibiotic residues, radioactive substances, bovine spongiform encephalopathy, allergies, and cloned beef. Respondents were asked to rate them on a scale from

Table 18.1 Summary of the web-based surveys

		1st		2nd		3rd (4th following)				5th (6th following)			
Period		Oct–Nov 2011		Mar 2012		Jan 2013		Mar 2013		Feb 2014		Mar 2014	
Company		Nikkei Research Inc.				Nihon Research Center							
		N	%	N	%	N	%	N	%	N	%	N	%
Total		4363	100.0	5028	100.0	6357	100.0	1881	100.0	9678	100.0	1822	100.0
Sex	Male	2165	49.6	2641	52.5	3385	53.2	962	51.1	5169	53.4	953	52.3
	Female	2198	50.4	2387	47.5	2972	46.8	919	48.9	4509	46.6	869	47.7
Age group	20 ~ 29	882	20.2	873	17.4	936	14.7	344	18.3	1101	11.4	357	19.6
	30 ~ 39	839	19.2	1014	20.2	1485	23.4	410	21.8	2074	21.4	384	21.1
	40 ~ 49	864	19.8	1078	21.4	1550	24.4	412	21.9	2542	26.3	388	21.3
	50 ~ 59	861	19.7	1047	20.8	1428	22.5	384	20.4	2364	24.4	355	19.5
	60 ~ 69	917	21.0	1016	20.2	958	15.1	331	17.6	1597	16.5	338	18.6
Residential area	Hokkaido	70	1.6	95	1.9	191	3.0	124	6.6	387	4.0	83	4.6
	Tohoku	433	9.9	582	11.6	775	12.2	203	10.8	884	9.1	186	10.2
	Kanto	1462	33.5	1255	25.0	1278	20.1	236	12.5	2617	27.0	359	19.7
	Hokuriku	281	6.4	382	7.6	480	7.6	182	9.7	536	5.5	87	4.8
	Chubu	347	8.0	516	10.3	796	12.5	276	14.7	1607	16.6	326	17.9
	Kinki	500	11.5	624	12.4	923	14.5	227	12.1	1469	15.2	346	19.0
	Chugoku	367	8.4	479	9.5	621	9.8	233	12.4	795	8.2	144	7.9
	Shikoku	292	6.7	383	7.6	405	6.4	192	10.2	392	4.1	88	4.8
	kyushu	611	14	712	14	873	14	208	11	991	10	203	11

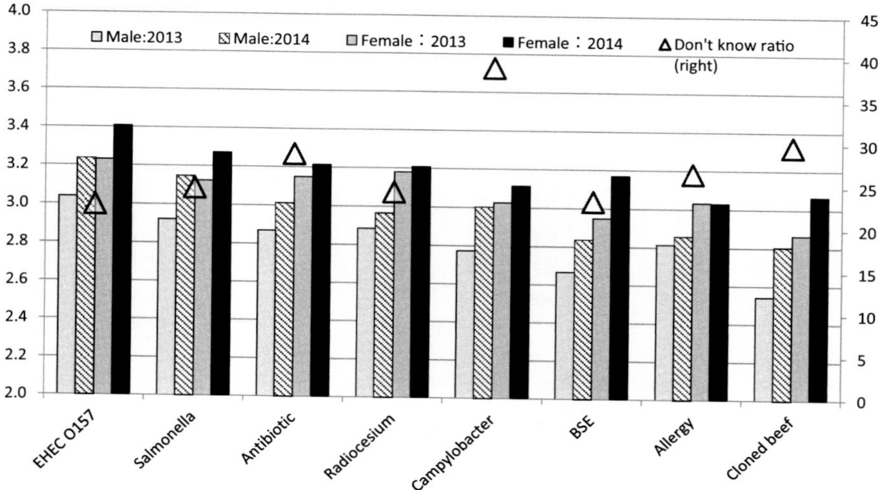

Fig. 18.1 Risk perception and the rate of choosing the "don't know" option. *Left axis* shows the average of responses from no risk (0) to high risk (5) and *right axis* shows the percentage of responses choosing "don't know" option

no risk (0) to very high risk (5). Considering cases having no idea about hazards or no ability to assume risk, the option of "have no idea" was included. Average perceived risk levels by gender for the 3rd and 5th surveys are shown in Fig. 18.1.

Compared to the 3rd survey, the 5th survey showed a general tendency for higher risk perception. Women were more apt to assume higher risk in all hazards than men. For individual hazards, enterohemorrhagic *E. coli* was ranked highest in every survey and for both genders. Approximately 25 % of respondents chose "Don't know" to each hazard, with around 40 % in the case of *Campylobacter*. Radiocesium was ranked 3rd highest among women while 5th highest among men.

To investigate the perceived risk level of Japanese beef compared to imported beef, in the 6th survey we asked the risk level of three hazards from eating beef distributed in US, France, and China: enterohemorrhagic *E. coli*, bovine spongiform encephalopathy, and radiocesium. Half of the respondents were asked to answer the Japanese beef risk level at first and the others were asked Japanese beef risk level at the end while the other country of origin was randomly presented. Figure 18.2 shows the results from these questions. The perceived risk level of acquiring enterohemorrhagic *E. coli* or bovine spongiform encephalopathy by eating Japanese beef were lower than the other countries, whereas radiocesium was similar to USA and France and lower compared to Chinese beef. In the cases where Japanese beef was presented at the end, the perceived risk level was lower than when it was presented first.

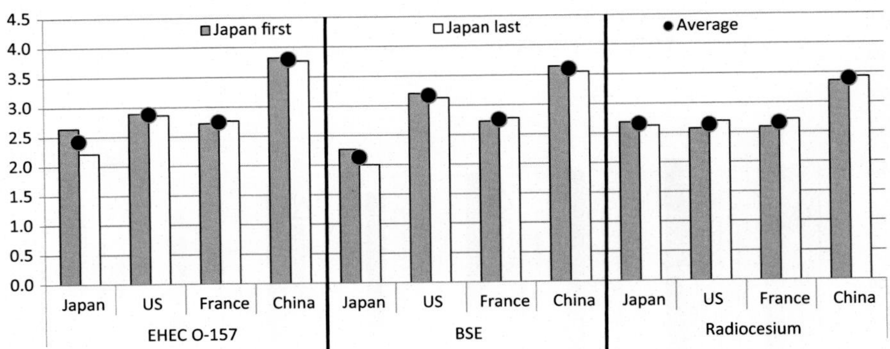

Fig. 18.2 Comparative risk perception of beef distributed in Japan, USA, France, and China. This figure shows the average of responses from no risk (0) to high risk (5)

18.3.2 Attitude and Willingness to Pay for Foods from the Disaster-Affected Area

Attitudes toward foods produced around the disaster-affected area were asked in the 2nd, 3rd, and 5th surveys. We provided several statements and the respondents were asked to answer to what extent they agreed with the statements; 4 and 6 Likert-scales were used in the 3rd and 5th surveys, respectively. The ratios of those who agreed to each statement are shown in Fig. 18.3. The results showed that approximately half of the respondents somewhat agreed to the purchase of foods away from the devastated area after the nuclear power plant explosion. Approximately 45 % of respondents thought agricultural produce from Fukushima should not be used for a school lunch. These ratios slightly increased over the last 3 years. In contrast, approximately 50 % were willing to purchase food from the Kanto/Tohoku region as well as Fukushima prefecture to support the recovery and the ratio was increasing. Approximately 75 % of responders considered that they can contribute to the recovery by purchasing food from the devastated area.

Willingness to pay for foods produced around the radiation-affected area is shown in Fig. 18.4. Respondents were asked to choose the highest price they would pay for an item of food if radiocesium was not detected or was below the regulation level from 0 % (do not want to buy) to 200 % (twice as much as normal prices) in 10 % increments. In all surveys, approximately 70 % of the responders did not want to pay the normal price if radiocesium was below the regulation level. In addition, the ratio of those who did not want to buy (0 %) increased between the 2nd and 3rd surveys from around 10 to over 20 %. Similarly, even if radiocesium was undetected in the produce, around 15 % of responders answered that they did not want to buy it in the 3rd and 5th survey, which was an increase of more than 5 % from the 1st and 2nd surveys. Therefore, although the radiocesium regulation levels in general foods was strengthened in April 2012 from 500 to 100 Bq/kg, the survey results indicate that this revision did not increase consumer confidence about the safety of foods produced around the devastated area.

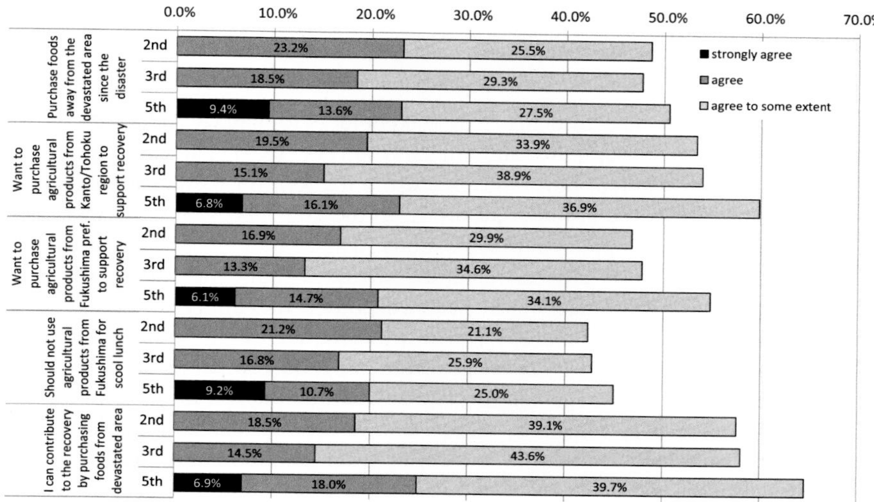

Fig. 18.3 The ratio of those who agree to each statement

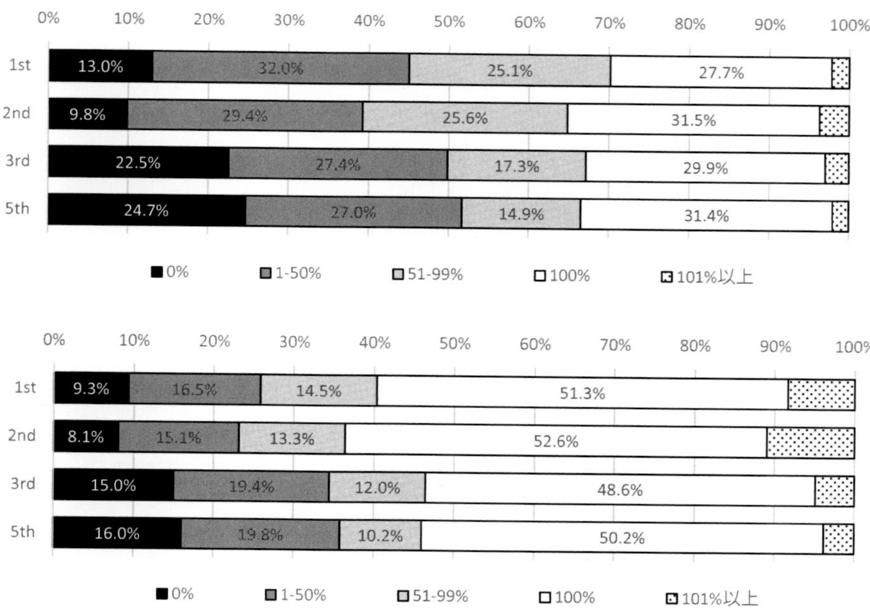

Fig. 18.4 WTP for foods produced in the disaster affected area. Radiocesium at below the regulation level (*upper results*) and not detected (*lower results*)

18.3.3 Trust About Risk Management of Radioactive Substances in Foods

Consumer trust about the efforts made by stakeholders to manage radiation risk in foods was queried in the 2nd, 3rd, and 5th surveys (Fig. 18.5). The choices ranged from (strongly) agree to (strongly) disagree using the 4 or 6 Likert-scale, which is similar to that used for the attitude questions. In the 2nd survey, slightly less than 20 % of the respondents agreed to the questions "The central government provides necessary information for citizens enabling them to judge the safety of food regarding radioactive substances" and "I can trust on the radiation risk management taken by the central government." The affirmative answer to trust about risk management measures taken by local government, food companies, and retailers ranked higher than risk management measures taken by the central government in the 2nd survey. The percentage of "agree" and "somewhat agree" regarding trust about radiation risk management of central or local government as well as other stakeholders in the food chain increased in the 3rd and 5th surveys. However, the results showed that more than half of the respondents "(somewhat) disagree" about the trustworthiness of radiation risk management in foods 3 years after the incident. Approximately 50 % of respondents answered "agree" to the question "In order to improve safety, the stricter the standard value of radioactive substances in food, the better" in the 2nd and 3rd survey and the percentage increased to over 60 % in the 5th survey. This indicates a greater consumer desire for stricter regulations about radioactive substances in foods as time passes.

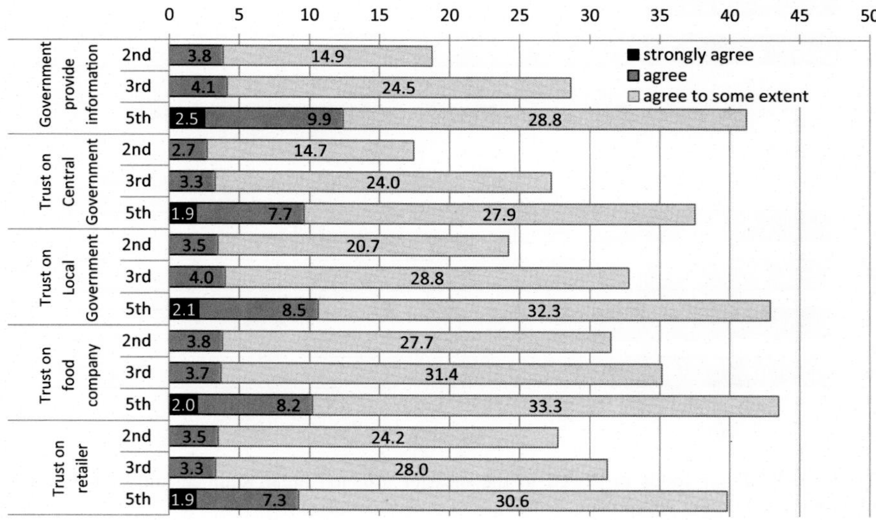

Fig. 18.5 Trust on the radiation risk management body

Table 18.2 The correct answer rate for the questionnaire about radiation risk and risk management measures

	2nd (%)	3rd (%)	5th (%)
Meaning of Becquerel and Sievert	35.9	41.0	37.1
Biological half life of radiocesium	29.2	33.5	12.3
Health effect of low dose exposure	14.7	14.5	22.3
Permissible dose level	15.2	14.5	11.7
Natural exposure dose in pease time	27.6	26.7	
Repair function of gene	27.7	23.8	
Once shipment ban applied, it take 1 month before reshipping	11.0	11.8	
Standard value of radiocesium		7.8	12.8
Current radiocesium exposure level in Fukushima			18.9
Inspection result of beef			19.3

18.3.4 Knowledge About Radiation Risks and Risk Management Measures

To investigate the knowledge level of the respondents, we showed several correct/incorrect descriptions about radiation risk and risk management measures. Respondents were asked to answer if the description is "true," "not true," or "I don't know". The presented descriptions were not the same between the surveys due to the changing conditions and provided information. Table 18.2 shows the correct answer rate before information was provided. The result showed that approximately 40 % of respondents understood the meaning of Becquerel and Sievert. Those who knew the difference between biological half-life and physical half-life of radiocesium was approximately 30 % in the 2nd and 3rd survey, while it was 12.3 % in the 5th survey. Natural exposure dosage in Japan and repair function of genes were queried in the 2nd and 3rd surveys and approximately 25 % of people were aware of them. Knowledge of the permissible dose from foods (1 mSv), standard value of radiocesium in general foods (100 Bq/kg) and current radiocesium exposure levels in Fukushima were lower, less than 20 % in the 5th survey.

18.3.5 Satisfaction Levels for Radioactive Substance Management in Food

We asked about satisfaction with radioactive substance management in food by the government in the 3rd and 5th surveys. Satisfaction levels by gender and age group are shown in Fig. 18.6. The percentage of respondents who answered "not satisfied" was 19.3 and 29.3 % in the 3rd and 5th surveys respectively, with a tendency to increase with age. Although trust levels increased as time passed, satisfaction

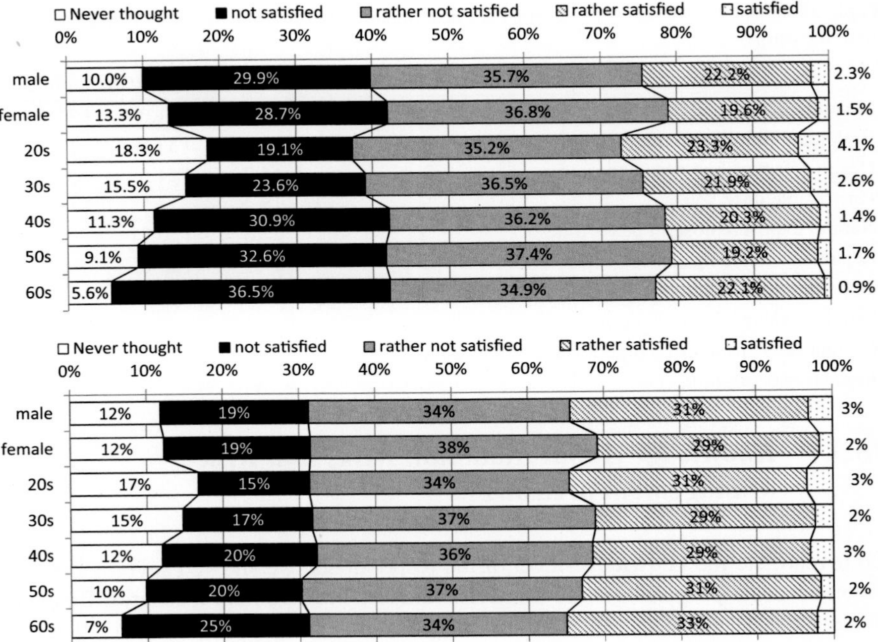

Fig. 18.6 Satisfaction level of radiation risk management in the 3rd (*upper results*) and 5th (*lower results*) surveys

decreased. Only a few percentage of respondents answered "satisfied." The younger respondents were more likely to respond "have never thought about it."

The relationship between satisfaction levels for radioactive substance management in food and WTP for food from disaster-affected areas is shown in Fig. 18.7. In the 3rd survey, as satisfaction levels rose, willingness to pay for foods from the devastated area increased. The gap between "below regulation level" and "not detected" became smaller as the satisfaction level increased. However, the average WTP among those satisfied with the management was not highest in the 5th survey. The relationship between knowledge and risk perception was inexplicable in the 5th survey (Fig. 18.8). Higher knowledge and lower risk perception related to satisfaction in the 3rd survey; however, the correct answer rate of "satisfied" people was higher, but perceived risk level was not lowest in the 5th survey. Further research to investigate the constituent of satisfaction is expected.

18.3.6 Results of the Beef Choice Experiment

Table 18.3 shows the estimated parameters applied to the multinomial logit model of a beef choice experiment. Before the experiment, we provided a 3-min movie

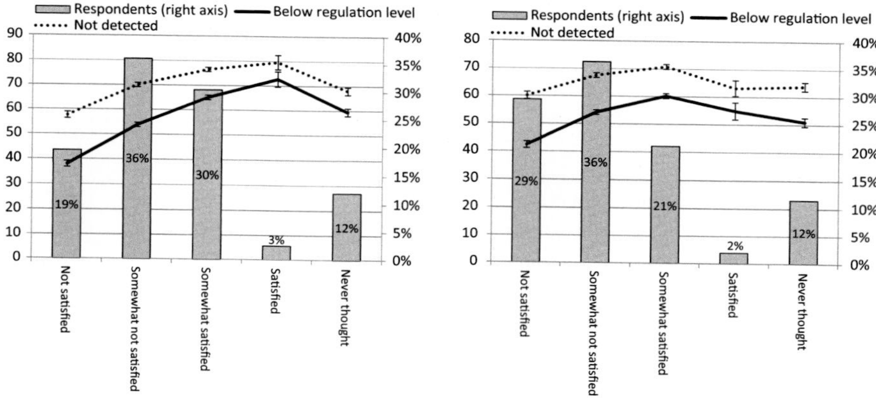

Fig. 18.7 Relationship between satisfaction level and WTP for foods from the devastated area

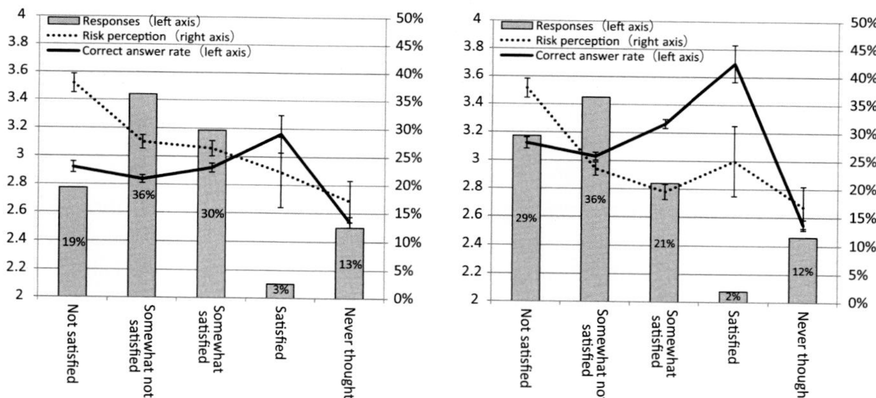

Fig. 18.8 Relationship between satisfaction level, risk perception, and correct answer rate

explaining the risk and risk management of radioactive substances in foods. However, 33.4 % of those who were supposed to be provided with information stopped watching without waiting for the end. Therefore, only approximately 30 % of respondents were provided with information. The attributes for beef were origin of beef (four domestic sites in Japan, USA and Australia), price (from 78 JPY to 468 JPY per 100 g) and inspection results of radioactive substances. Inspection results were shown in "Becquerel/kg in number" (less than 10 Bq/kg, less than 25 Bq/kg, less than 50 Bq/kg or less than 100 Bq/kg) or "Words" at a similar level as a number (not detected, less than 1/4 of regulation level, less than 1/2 of regulation level or below regulation level). Respondents were asked to answer nine times for choice experiments for which the same alternatives were presented in the 4th or 5th choice and 9th choice. Only those who chose the same alternative

Table 18.3 The estimated parameters applied to the multinominal logit model

		Model 1		Model 2	
		Coefficients	P value	Coefficients	P value
attribute	Constant : Domestic	0.253***	0.00	0.280***	0.00
	Constant : Import	0.280***	0.00	0.253***	0.00
	Price	−0.005	0.47	−0.005	0.46
	Production site : Hokkaido	0.032	0.22	0.032	0.22
	Production site : Fukushima	−0.029	0.13	−0.081*	0.10
	Production site : Kagoshima	−0.001	0.98	0.000	0.99
	Production site : US	−0.029	0.28	−0.028	0.29
	Contamination level: in Bq/kg	0.002	0.94	0.003	0.94
	Contamination level: in words	0.028	0.38	0.028	0.37
Cross term	Fukushima* Complete info.	0.102*	0.06	0.095*	0.08
	Fukushima* previous knowledge			0.218**	0.02
	Fukushima* risk perception			−0.061***	0.00
	Fukushima* risk can't judge			−0.173**	0.02
	Fukushima* satisfaction level			0.096***	0.00
	No. of observation	29,832		29,832	
	AIC/N	2.762		2.761	

* ($p<0.1$), ** ($p<0.05$), *** ($P<0.01$)

in these two choice cases were applied to the analysis. Moreover, those who answered "Don't know" to some questions were excluded from the analysis. Therefore, a total of 29,832 samples were utilized for this analysis. Table 18.3 shows the estimated parameters used in the multinomial logit model.

Both models 1 and 2 show negative signs toward the production site "Fukushima" while the parameters for radiation contamination levels were not significant. This indicates the possibility that consumers focus more on the production site rather than on the result of a radiation inspection. Information was effective at increasing acceptance of beef produced in Fukushima. Also, the knowledge improved WTP for Fukushima beef. Perceived risk level also affected acceptance of Fukushima beef and those who answered "I can't judge the risk level" negatively evaluated Fukushima beef. In the previous descriptive analysis the relationship between WTP and satisfaction level were ambiguous; however, the estimated parameter of this choice experiment indicated recovered price evaluation from Fukushima labeling among those who satisfied with radiation risk management.

18.3.7 Donating Behaviors for Devastated Area and Food Safety Risk Management

Web-based donation experiments were implemented in the 4th ($N=1881$) and 6th ($N=1822$) surveys. Participants were selected from the 3rd and 5th surveys, respectively, considering the age group, sex, residential area, and WTP. Before participating in this game, they were asked how much they would be willing to donate if they acquired 10,000 JPY in the 6th survey, which was not asked in the 4th survey. After confirming their intention to participate in the experiment, participants received 300–1300 JPY as a result of a two-player Ultimatum Game. Next, we asked their intention to draw a lottery or not using the money they received in the Ultimatum Game. If they participated in the lottery, they would get 100–10,000 JPY and if they refused to participate, they would get the money they had acquired at that stage. The average amount of money acquired at this stage was 821.0 JPY in the 4th and 628.3 JPY in the 6th survey.

Many contributions were collected after the Great East Japan Earthquake to support affected people and recovery activities. In our research, we presented 10 activities and asked each participant to donate as much as they wanted using the money they received from the above experiment. Their donating behavior is shown in Tables 18.4 and 18.5. Women donated more than men, and the donation rate increased in the older groups. Overall donation rates increased in the 6th survey (54.4 %) compared to the 4th survey (46.5 %). The greatest amount, approximately

Table 18.4 Donated rate by sex and age group

	Donation rate		
	4th		6th
	Actual (%)		Projected (%)
Male	47.1	55.5	58.8
Female	52.0	58.0	62.4
20s	31.7	43.7	50.5
30s	43.0	49.9	55.8
40s	53.3	58.8	60.2
50s	57.0	63.2	65.5
60s	63.7	68.4	71.3

Table 18.5 Donated rate by activities

Activities	4th (%)	6th (%)
Recovery of Tsunami affected area	15.1	15.4
Health monitoring for radiation exposed children	5.4	7.8
Compensation for farmers in Fukushima	5.1	4.7
Research on development of new energy	4.8	6.4
Research on decontamination of radioactive substances	4.3	5.4
Research on health effect of radiation exposure	4.3	5.7
Compensation for farmers outside Fukushima	2.2	3.0
Radiation monitoring of marine products	1.9	2.1
Radiation monitoring of animal products	1.8	2.4
Other food radiation monitoring	1.7	1.8
Total donation	46.5	54.4
Respondents received	53.5	45.6

30 % of all donations, was contributed to the recovery of tsunami-affected areas, followed by health monitoring of children from radiation exposure. In the 6th survey, more participants contributed to research such as recyclable energy, health effects of low dose exposure, and decontamination of radioactive substances. The difference in the donated amount between compensation for farmers in Fukushima and outside Fukushima decreased. Total donations for intensification of food monitoring was 5.4 % (4th) and 6.2 % (6th), which were the 2nd and 4th largest donations, respectively.

18.3.8 Consumer Requirement for Food Safety Measures

In the 6th survey, we asked the respondents to select and rank food safety measures that they considered important. We displayed 18 options randomly ordered to avoid the ordering effect; options shown at the top were usually selected more often. As a result, communicating risk with consumers about radiation risk and food poisoning were highly prioritized, whereas communicating risk about bovine spongiform encephalopathy with consumers was not highly ranked (Table 18.6). Following

Table 18.6 Consumer demand for food safety measures

	1st (%)	2nd (%)	3rd (%)	4th (%)	5th (%)
Risk communication for consumers (radioactive substances)	18.1	7.5	6.0	4.0	5.1
Risk communication for consumers (food poisoning)	9.8	14.5	4.7	4.5	5.7
Decontamination of soil	9.8	5.5	4.9	7.7	7.1
Risk communication for food industry (food hygiene)	8.7	7.8	10.6	4.9	5.1
Enhance radio Cs inspection (animal products)	7.8	9.3	9.1	9.3	7.8
Implement BSE blanket inspection	6.5	4.7	6.8	7.9	6.1
Enhance radio Cs inspection (marine products)	6.4	11.7	9.8	8.9	5.5
Research on health effect of low dose radiation exposure	4.7	3.9	3.7	3.1	7.4
Enhance restaurant hygiene monitoring	4.2	3.7	4.3	3.4	5.4
Develop BSE vaccine	4.2	2.5	2.8	2.6	4.5
Hygiene control in food distribution	3.9	6.5	7.1	6.9	7.2
Research on hestimating radiation exposure from food intake	3.3	3.5	2.4	3.9	5.2
Hygiene control in restaurants	2.8	4.2	3.3	5.3	5.4
Enhance radio Cs monitoring (plants)	2.3	3.9	9.8	8.3	6.3
Develop technology for edible raw beef	2.2	1.9	2.7	3.7	4.0
Risk communication for consumers (BSE)	2.1	3.3	4.7	7.2	3.5
Enhance slaughterhouse hygiene	1.8	3.2	4.1	5.4	4.2
Enhance farm hygiene	1.4	2.1	3.1	3.1	4.6

these options, consumers thought that radiocesium inspection in food, especially animal and marine products should be enhanced.

18.4 Discussion and Conclusions

Serious efforts to reconstruct the devastated area have been undertaken since the earthquake on March 11, 2011. Regarding the radiation contamination of foods, almost all marketed foods have been shown to be below the detection limit except for some seafood, fungi, wild vegetables, and wild animals. This series of surveys indicates that consumers' perceived risk level of radioactive substances in foods is not high compared with the microbial risks. Moreover, risk from radiocesium food contamination in Japan was considered to be at a similar level to that in USA and France and lower than foods distributed in China. Meanwhile, risk of bovine spongiform encephalopathy and *E. coli* O157 in Japan were regarded as the lowest among the four countries. However, we could argue that consumer anxiety is not reduced if we consider the results of consumer awareness and WTP for food from the disaster-affected area.

Trust about risk management taken by stakeholders, including the government and food industry, was shown to be recovering. It seems contradictory, but consumer satisfaction did not recover like trust. Moreover, there was a correlative relationship between satisfaction level and lower risk perception as well as higher WTP for foods from the disaster-affected area in the 3rd survey; although the relationship weakened in the 5th survey. Perceived radiation risk from foods was not ranked lowest and WTP was not ranked highest among those who revealed highest satisfaction levels. Albeit further research is needed to clarify this point, one explanation is that they are satisfied with being able to choose foods that originate away from the disaster affected area by checking the production site on the label. This is partially indicated by that most people do not know about the standard value of radiocesium and/or the permissible level of exposure from foods as well as current exposure level in Fukushima. A similar result also was obtained from the beef choice experiment, where the estimated coefficient of production area as "Fukushima" was negative but the radiation contamination level (results of inspection) was not significant.

Consumers are carrying an additional burden of collecting information to understand the effects of radioactive substances on health. It is easier to feel secure by choosing foods that originate away from the disaster site rather than trying to understand risk management for radiocesium contamination and the current contamination situation. Therefore, it makes sense for consumers to avoid risk by selecting food from areas as far away from the incident site as possible.

Consumer requests for communication about radiation risk are strong as indicated in the 6th survey. To reduce the cost of collecting and processing public information, communication content and methods should be developed by involving various entities, including the food industry, experts, and consumer

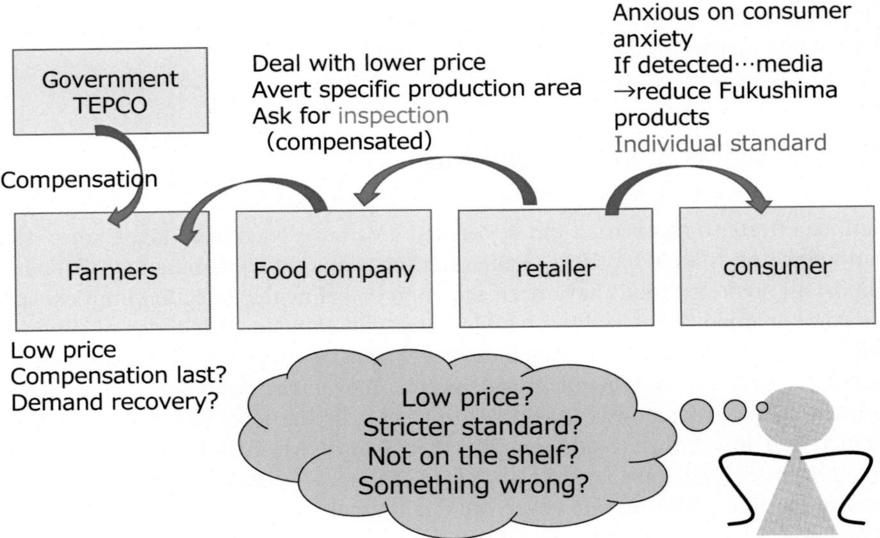

Fig. 18.9 Vicious circle on food market in devastated area

organizations. The media such as TV, newspapers, and magazines play a major role by providing information to citizens. However, less information about the situation in the devastated area is reported in the media as time passes. When radiocesium inspection results indicate values in excess of the standard value, then it is usually reported in the media, whereas a normal situation is usually not reported.

We might wait for consumers to forget about the hazard or risk, but it might take longer time. In addition, because consumer knowledge on risk and risk management measures are limited and focus more on production site rather than inspection result, it has the potential to lead consumers to regard a particular region as dangerous in the long term (Fig. 18.9), or indicates a possibility to occur regime shift as Yoshida has mentioned.

We consider that it is preferable for Japanese consumers to feel secure in the long term by acquiring knowledge about radiation risk, even it requires effort. Otherwise, they will be faced with some anxiety with every food purchase or they will be unable to enjoy the food culture or special local products produced in the affected area. Moreover, reconstruction of agricultural production in Tohoku area, which is an important food supply base for the metropolitan area, is crucial when considering the future stability of Japan's food supply. Consumer support is a vital element for promoting the reconstruction. The food risk we are facing is not only caused by radioactive substances. In order to enjoy a healthy diet and a healthy life, we should consider other food safety risks, such as the availability of food (food security), ecosystem sustainability, natural resources, and local infrastructure.

References

Food Safety Commission of Japan (2013) Food safety monitor reports (August 2013, in Japanese) https://www.fsc.go.jp/monitor/monitor_report.html

Hangui S (2013) Consumer's response to the information provided about contamination of the food by Radioactive Materials. J Agric Rural Econ 85(3):173–180, (in Japanese)

Kito Y (2012) How do lay people perceive the health effect of radioactive substances?: an analysis of internet survey data, Y. Kito. Agric Econ 78(1) (in Japanese)

Kurihara S, Ishida T, Maruyama A, Matsuoka N, Sugawara S (2013) Construction of the system for releasing the result of radiation test and Consumer's preference analysis. J Food Syst Res 20 (3):209–214 (in Japanese)

Ujiie K (2012) Consumer's evaluation on radioactive contamination of agricultural products in Japan: decomposition of WTA into a part due to radioactive contamination and a part due to area of origin. Food Syst Res 19(2):142–155 (in Japanese)

Ujiie K (2013) Transition of consumers evaluation on radioactive contamination of agricultural products in Japan. J Rural Econ 85(3):164–172 (in Japanese)

Yoshida K (2013) An econometric analysis of consumer's averting behavior caused by the radioactive contamination of agricultural, forest and fishery products. J Rural Econ:258–265, (in Japanese)

Chapter 19
Imaging Techniques for Radiocesium in Soil and Plants

Ryohei Sugita, Atsushi Hirose, Natsuko I. Kobayashi, Keitaro Tanoi, and Tomoko M. Nakanishi

Abstract Various radioisotope imaging techniques have been used at the Graduate School of Agricultural and Life Sciences, University of Tokyo, to analyze samples containing radiocesium (^{137}Cs and ^{134}Cs). There are two types of samples: (1) environmental samples contaminated by the fallout from the Fukushima Daiichi nuclear power plant accident, which contain relatively low concentrations of radiocesium and (2) laboratory samples from tracer experiments conducted at the radioisotope institution containing relatively high concentrations of ^{137}Cs. The first technique used to visualize radiocesium in soil and plants was radioluminography (RLG). RLG, which makes use of an imaging plate, has a dynamic range that is large enough to detect both environmental and tracer-added samples. To quantify radiocesium distributions, the samples were frozen and sliced before contact with the imaging plate. This freezing procedure after sampling is for preventing radiocesium movement during slicing and measurement of ^{137}Cs distribution. After slicing, two detection methods were employed: RLG and microautoradiography (MAR). MAR is the conventional and older method for imaging radioisotopes based on the daguerreotype process. We applied this method to frozen sections and obtained ^{137}Cs distributions at a higher resolution than with RLG. Following this, we employed a non-destructive method for imaging ^{137}Cs movement in a living plant. We developed the visualization technique called real-time radioisotope imaging system and then demonstrated ^{137}Cs movement from soil to rice plants using a chamber containing paddy soil, water, and rice plants. Lastly, ^{42}K obtained by ^{42}Ar–^{42}K generation enabled a comparison between the movement of ^{137}Cs and ^{42}K. The mechanism of Cs transport has been reported to have some relationship with the K transport system, so experiments using both ^{137}Cs and ^{42}K would be useful for clarifying the mechanism in more detail.

Keywords Fukushima Daiichi nuclear power plant accident • Imaging plate • Live imaging • Microautoradiography • Radiocesium • Radioisotope imaging • Radioluminography • Real-time radioisotope imaging • Soil • Whole-plant imaging • ^{137}Cs distribution

R. Sugita • A. Hirose • N.I. Kobayashi • K. Tanoi (✉) • T.M. Nakanishi
Graduate School of Agricultural and Life Sciences, The University of Tokyo, 1-1-1 Yayoi, Bunkyo-ku, Tokyo 113-8657, Japan
e-mail: uktanoi@mail.ecc.u-tokyo.ac.jp

T.M. Nakanishi, K. Tanoi (eds.), *Agricultural Implications of the Fukushima Nuclear Accident*, DOI 10.1007/978-4-431-55828-6_19

19.1 Radioluminography

19.1.1 Radioluminography Using an Imaging Plate

Radioluminography (RLG) is a type of autoradiography (ARG) based on stimulated luminescence and is widely used to visualize radiation from samples. Before the RLG technique was established, ARG using X-ray film was employed for visualization, despite its difficulties in handling, low quantitativity, and low sensitivity. We used the imaging plate (IP) (Fuji Film) as a stimulable phosphor plate because it has the advantage of a wide dynamic range of determination, re-usability, and high sensitivity.

There are many techniques available for detecting ^{137}Cs in samples and for measuring gamma ray emissions. For example, the germanium semiconductor detector and sodium iodide scintillation counter are often used for general determinations. These techniques are more suitable for ^{137}Cs determination compared to ARG because ARG mainly detects the beta rays of ^{137}Cs, which are easily self-absorbed in thick samples Nevertheless, ARG using an IP (i.e., RLG) has been a useful technique for observing ^{137}Cs distribution in thin samples from fields in Fukushima or in samples from the radioisotope laboratory. RLG has been used to quantify ^{137}Cs and other nuclides in plant samples (Kanno et al. 2007; Sugita et al. 2014).

A sample is placed in contact with the IP for a period of time that is dependent on the radiation level in the sample. During sample exposure to the IP, radiation photostimulates the fluorescent material attached to the IP and a radiation image is recorded. The radiation image is then analyzed by a He–Ne laser-based imaging system. The photostimulated fluorescent material releases its fluorescence as it is scanned by the laser, and the fluorescence is sequentially detected and digitized by the imaging system. The intensity of the fluorescence is called the photostimulated luminescence value, which has a linear relationship with radiation intensity up to at least five orders of magnitude. Conveniently, the afterimage on the IP can be eliminated by white fluorescence; hence, the IP can be used repeatedly. The sensitivity of the IP was more than 10 times higher than X-ray film, even though the exposure time for the environmental samples was longer on the latter due to their low radiocesium activity. To detect low radiocesium activity, we performed the exposure inside a shield made of lead and copper at around 4 °C (Fig. 19.1). The shield is expected to prevent exposure to background radiation and the low temperature can reduce fading during the long exposure time (Suzuki et al. 1997).

Because the RLG technique is easy to handle and generic in life science research, many samples derived from the Fukushima Daiichi nuclear power plant accident have been examined, which include the following: leaves of Japanese Beech (*Fagus crenata*) and needles of Japanese Black Pine (*Pinus thunbergii*) (Koizumi et al. 2013); soil, bamboo, and mushroom (Niimura et al. 2014); branches and leaves of *Cryptomeria japonica*, branches and leaves of *Thujopsis dolabrata* var. hondae, branches and leaves of *Prunus percica*, branches and roots of *Morus*

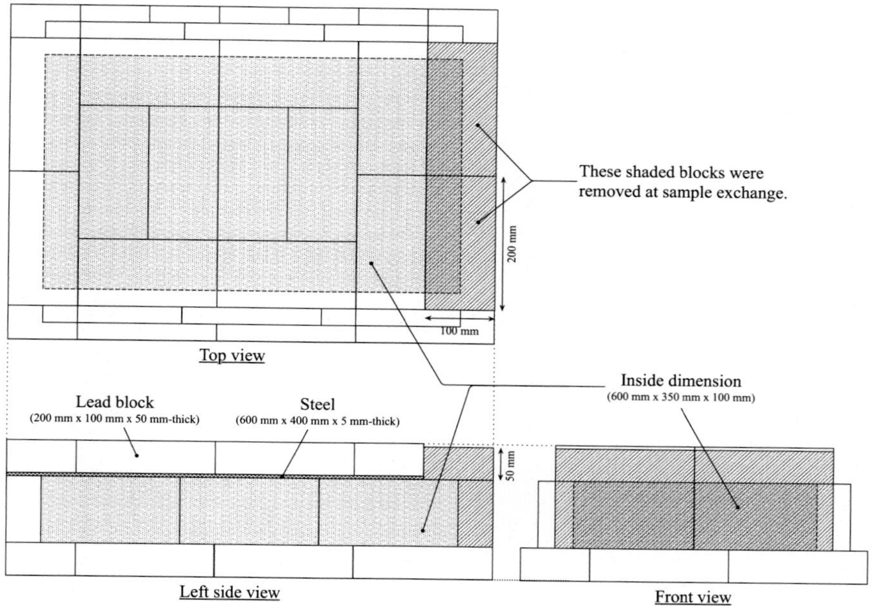

Fig. 19.1 The lead shield employed for detecting the low level of radiation

alba, lichen *Permotrema tinctorum*, and internodes and shoots of *Phyllostachys bambusoides* (Sakamoto et al. 2013); skeletal muscle of pig (Yamaguchi et al. 2013); green onion, carrot, and lettuce (Isobe et al. 2013); leaves of *C. japonica*, fallen leaves of *Quercus serrata*, and fallen leaves of *Pedicularis densiflora* (Tanaka et al. 2013); soil particles (Mukai et al. 2014); leaves of *Sasa palmate*, *Taxus cuspidate*, *Quercus serrate*, *Trifolium* spp., and *Equisetum arvense* (Mimura et al. 2014); leaves of cabbage and spinach (Shiba et al. 2013); leaves of bamboo and oak, and trunks of cedar tree and pine tree (Nakajima et al. 2014); leaves of beech, pieris, and oak (Furuta 2013); soil, plant, and dust samples (Itoh

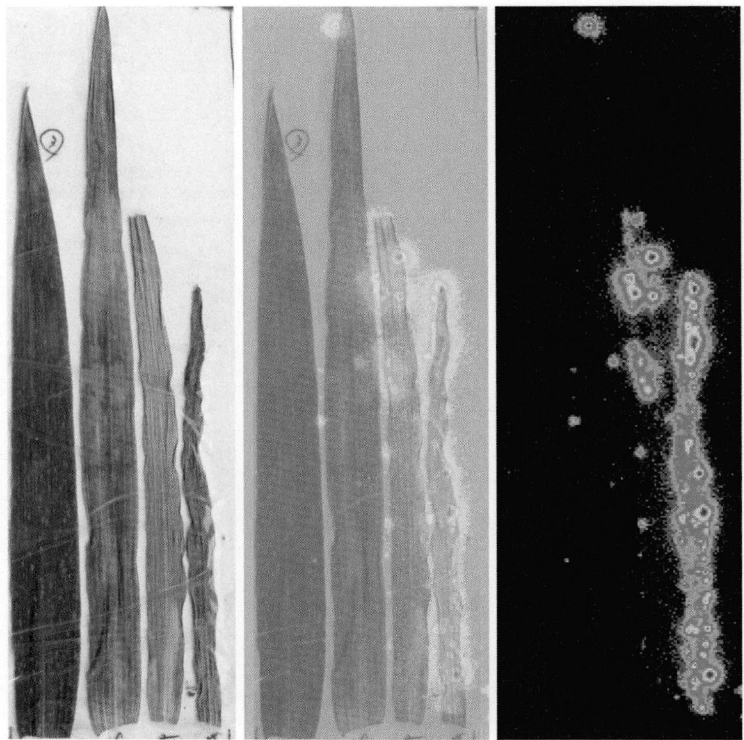

Fig. 19.2 The example of radiocesium image of barley leaves in May 2011 by radioluminography (RLG)

et al. 2014); filters (Zeissler et al. 2013); leaves and ears of wheat and barley (Tanoi et al. 2011; Tanoi 2013); soil, leaves of wheat, branches, leaves and fruits of peach, and rice plants (Nakanishi et al. 2012); bark of *Rhus vernicifera* DC. (Mori et al. 2012); bamboo (*Phyllostachys reticulate* (Rupr.) K. Koch) (Nakanishi et al. 2014); and branches of peach (Takata et al. 2012).

Most of the radioluminograms pointed out that direct contamination was observed as dots on leaves as well as on soils (Fig. 19.2), these dots having been observed as long as several years after the Fukushima Daiichi nuclear power plant accident. On the other hand, some images, such as rice plants (Nakanishi et al. 2012) and bamboo shoots (Nakanishi et al. 2014), showed a radiocesium distribution indicating possible (cesium) transportation inside/within the plant.

19.1.2 Radioluminography with Frozen Sections

The RLG images mentioned in the previous paragraph indicate how the contaminated nuclides were distributed in a sample at a centimeter or subcentimeter resolution (Fig. 19.3). To visualize the contamination in more details, it is necessary

The branch of the peach tree
was sampled in June 2011.

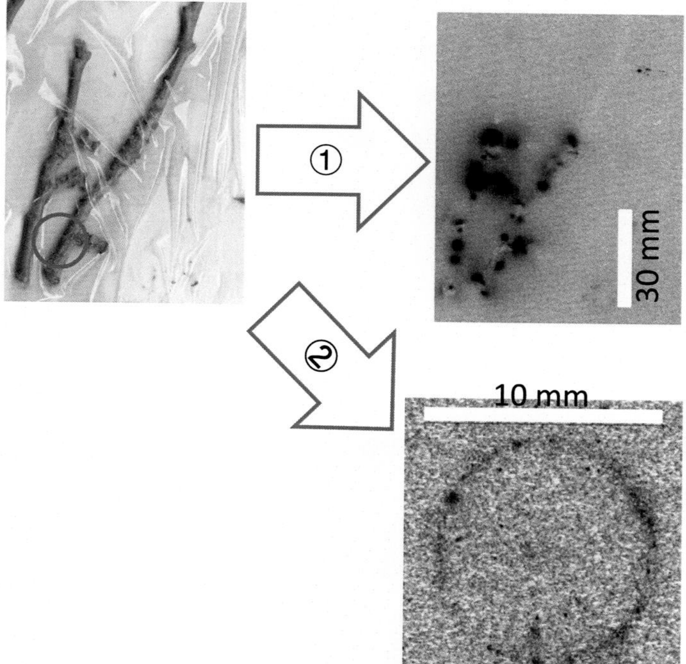

①The branch was exposed to an IP directly.
②The branch was kept under frozen condition. And the cross section of the branch was exposed to an IP.

Fig. 19.3 The different procedures between general RLG ① and RLG with frozen sections ②

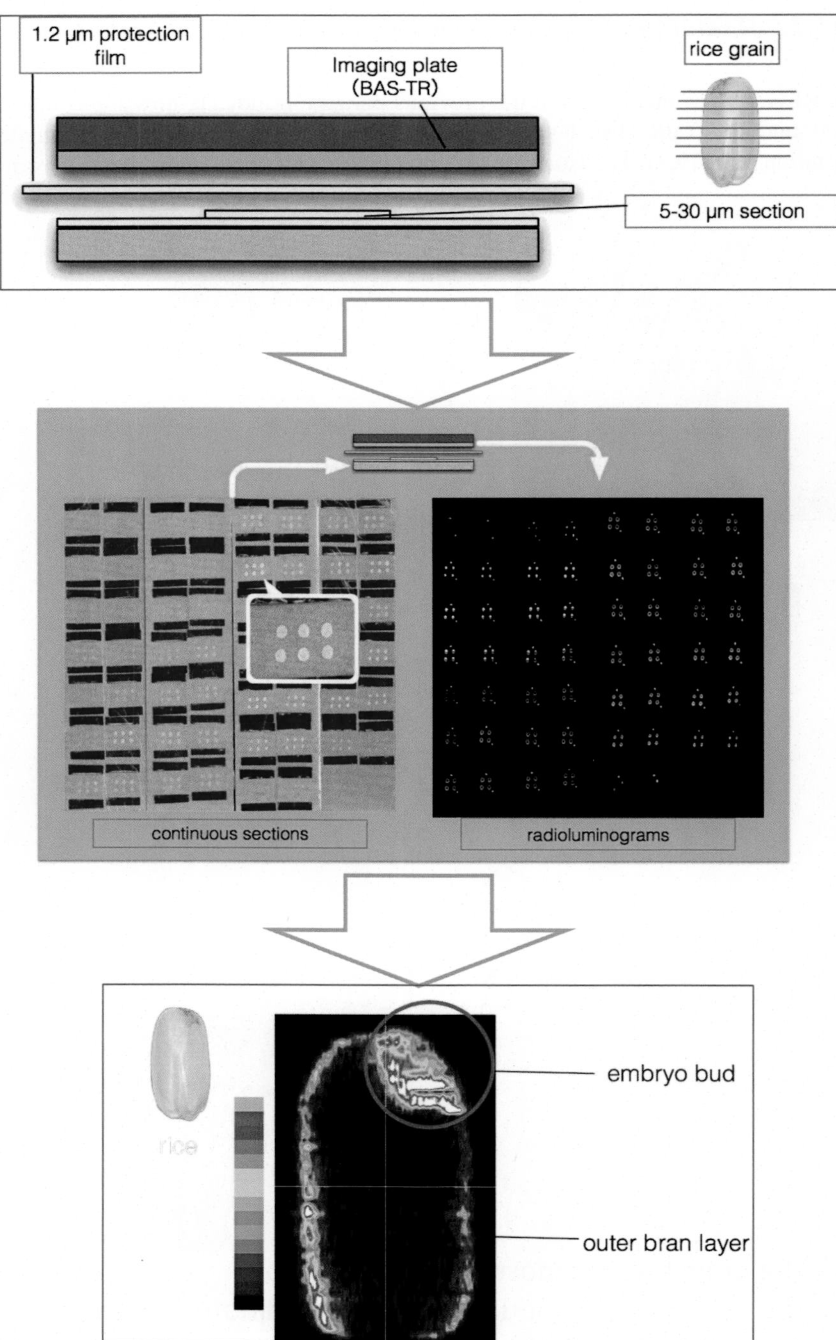

Fig. 19.4 The 3D construction from the sequential frozen sections in RLG. The sample shown here is a brown rice fed by ^{137}Cs

to place the sample closer to the IP; hence, flat sections are important for high definition RLG imaging. However, the ionic form of nuclides, especially radiocesium, can move inside the plant during sectioning, such as during fixation and dehydration. For these reasons, we performed RLG on frozen samples to diminish radionuclide movement. The samples were immediately frozen with liquid nitrogen after slicing and stored at −20 °C until RLG imaging. We observed degree of contamination in peach tree bark samples from 2011, but most of the environmental samples were less contaminated and it was difficult to detect radiation in the thin sections. In general, frozen section imaging by RLG was applied to a laboratory experiment where we controlled the radionuclide levels in the samples.

The good correlation between radionuclide activity and photostimulated luminescence value for RLG with an IP allows the construction of three-dimensional (3D) images using sequential sections. For example, a rice plant was supplied with ^{137}Cs via the roots several days after flowering; the brown rice was sampled when the rice grain was mature and subsequently embedded in resin under freezing conditions. Also, preparation of sequential sections and exposure to an IP were performed under freezing conditions. After contact, the IP was scanned with the analyzer to obtain the radioluminograms and the ^{137}Cs distribution was reconstructed in 3D using Image J software (Fig. 19.4). The ^{137}Cs tended to accumulate in the embryo bud and outer bran layer.

19.2 Microautoradiography

Microautoradiography (MAR) is based on a technique that was used to take photographs several decades ago. The distribution of radionuclides can be visualized in greater detail by MAR than with RLG, although RLG has a greater quantitative ability and sensitivity than MAR. For these reasons, MAR is only performed to obtain finer images and for higher resolution of radionuclide distribution.

It is necessary to freeze the sample for the MAR procedure, similarly to RLG with frozen sections so, we have developed the MAR technique using/under frozen conditions (Hirose et al. 2014). Using MAR, we observed/could observe the ^{137}Cs distribution in a rice grain at a sub-millimeter resolution (Fig. 19.5). The distribution of ^{137}Cs was revealed at the tissue level and showed low ^{137}Cs activity in the plumule and radicle. These details were unclear in the images using RLG (Fig. 19.5).

Figure 19.6 is a summary of the imaging techniques applicable to frozen sections. A suitable technique must be chosen to achieve the scientific aim.

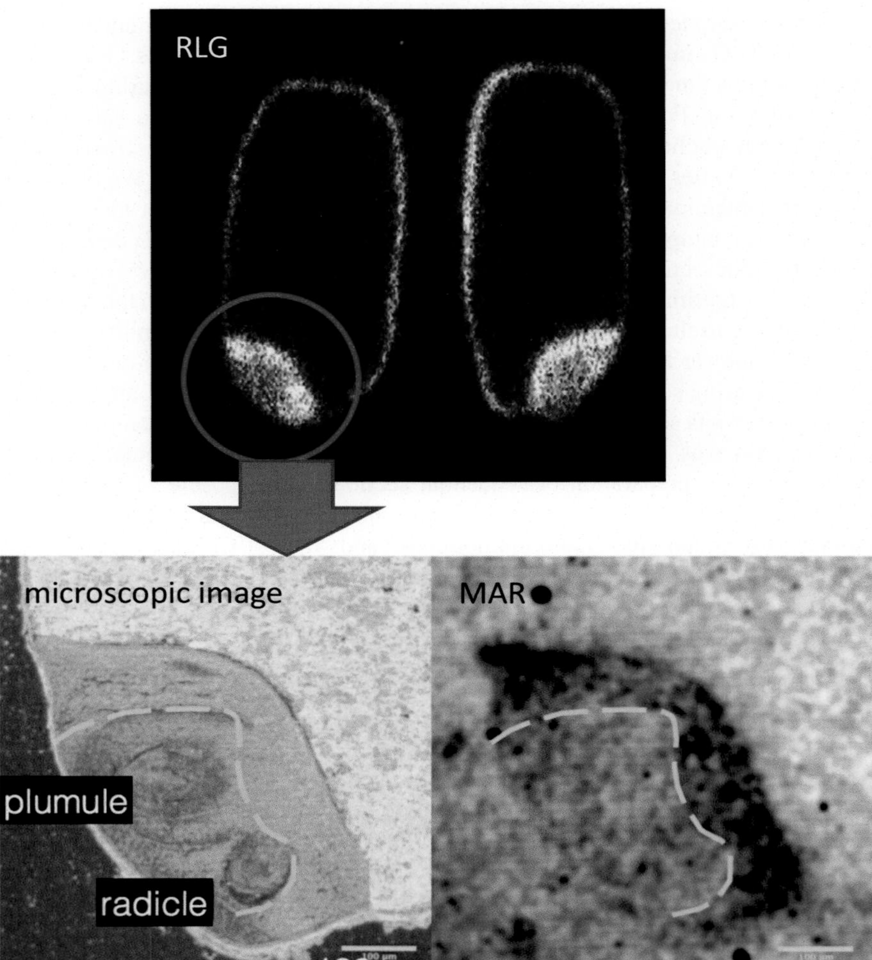

Fig. 19.5 The ^{137}Cs distribution in a rice grain visualized by RLG and MAR

19.3 Real-Time Radioisotope Imaging System

To understand the movement of solute in a plants, live imaging techniques are especially useful in such dynamic context. It is necessary for live imaging techniques to obtain data nondestructively. Radiation is a tool used to detect without destruction. Most of the live imaging systems using radiation or radionuclides are used in medicine, such as positron emission tomography (PET). There are several systems used to analyze plants using PET knowledge, such as the plant tomographic imaging system (PlanTIS) (Jahnke et al. 2009), or the positron emitting tracer imaging system (PETIS) (Fujimaki et al. 2010). In these systems, two annihilation

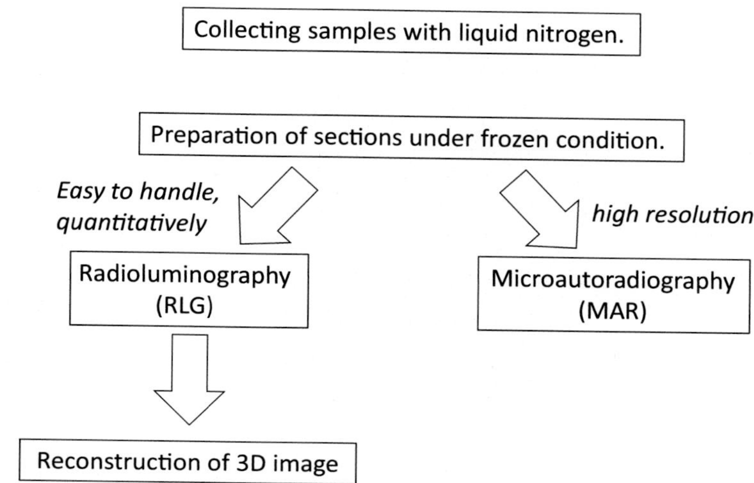

Fig. 19.6 Choice of the techniques to visualize the nuclides distribution of the sections

gamma rays, which are emitted from the positron emitting nuclide in a direction of 180°, are detected simultaneously so that the spatial distribution of the positron nuclide can be determined. Unfortunately, the half-lives of positron emitting nuclides tend to be too short to record the nuclides for several days, such as ^{11}C with a half-life of 20.39 min. Moreover, there are a lot of elements that are important in plant nutrition or harmful in the human diet, but have no available positron emitting isotopes.

We have developed a live imaging system to analyze plants over a long period with conventional radionuclides, called real-time radioisotope imaging system (RRIS) (Nakanishi et al. 2009). The RRIS can be applied to many kinds of nuclides available on the market because beta-rays and soft X-rays can be detected and the dynamic range of detection is about 1000 times (Sugita et al. 2014); ^{137}Cs, a beta-ray emitter, may be analyzed using this system.

19.3.1 How to Detect Radionuclides in the System

There are four steps to visualize the nuclide distribution in a plant sample (Fig. 19.7): (1) radiation is emitted from the radionuclides in a plant sample; (2) radiation is converted to photons (visible light), using a fiber optic plate with a scintillator (FOS); (3) visible light is converted to electrons, which are multiplied by a micro channel plate (MCP) with a GaAsP photoelectric surface; (4) electrons are detected using a charge-coupled device (CCD) camera. The image size is now 20 cm × 10 cm, and the resolution is around a few millimeters. To record the very

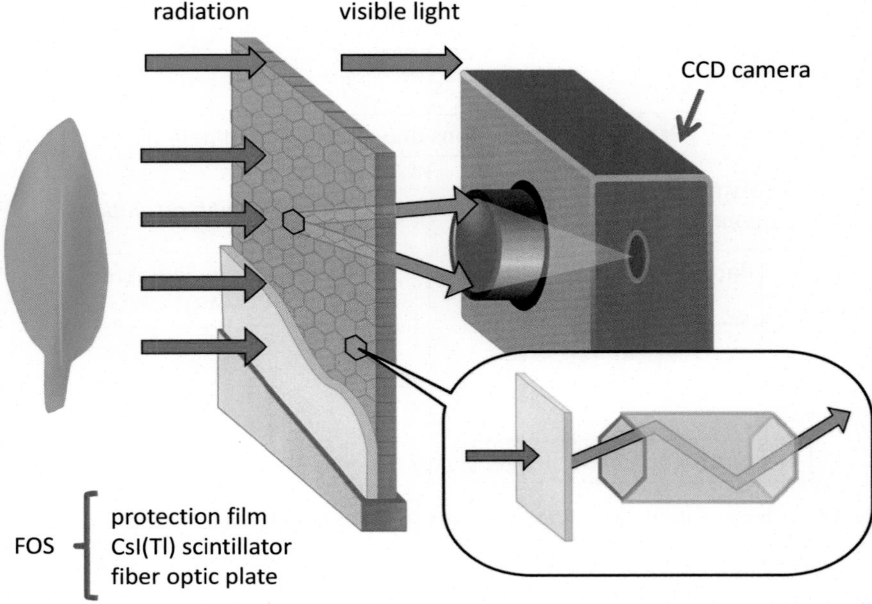

Fig. 19.7 How to make an image by real-time radioisotope imaging system. The radiation from the nuclides in a sample is converted to visible light by FOS containing scintillator. The CCD-camera catches the light sequentially

weak visible light, darkness is necessary between the scintillator and the camera. Thus, we used two types of plant chamber to illuminate the plant.

19.3.1.1 Type I: Shielding from Light-Emitting Diodelight

The plant was enclosed in an aluminum chamber containing light-emitting diode (LED) light to illuminate a plant leaf (Fig. 19.8 upper; (Yamawaki et al. 2010)). The thickness of aluminum between the plant and the FOS is 50–100 μm, which is enough to shield the visible light, but X-rays and relatively strong beta-rays can penetrate through the 100-μm aluminum film. The chamber is set inside the dark box and the CCD camera records the photons released from the FOS. The Type I system can only be applied to nuclides that emit X-rays or relatively strong beta-rays.

19.3.1.2 Type II: Intermittent Lighting System

An intermittent lighting system is employed for the RRIS, whereby the LED light is on when the photon counting camera is off, and *vice versa* (Fig. 19.8 bottom). The

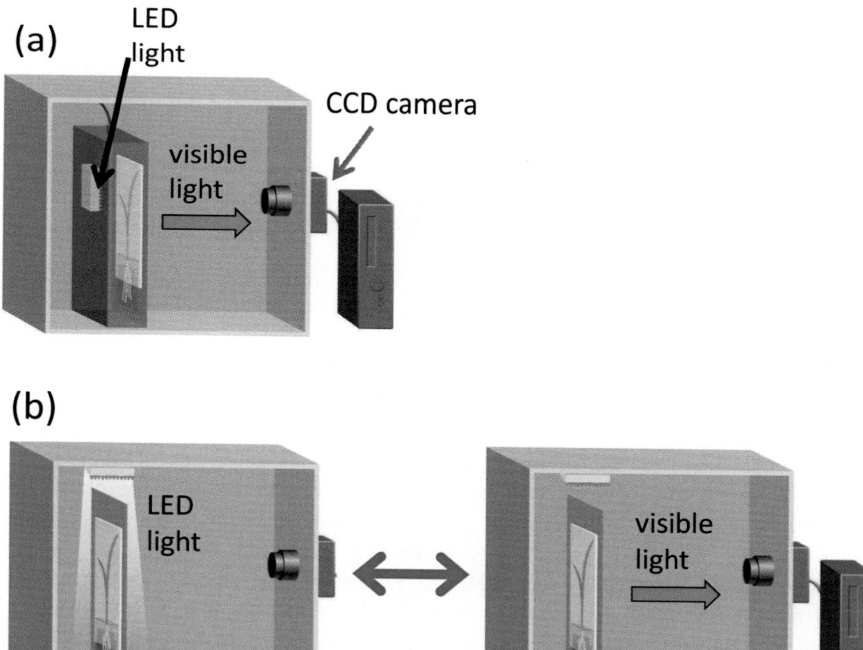

Fig. 19.8 Two types of chamber employed in the RRIS. (**a**) The chamber is covered with aluminum to prevent the LED light to leak to the CCD camera. (**b**) The chamber is open and the CCD light turns on only when the CCD camera is off

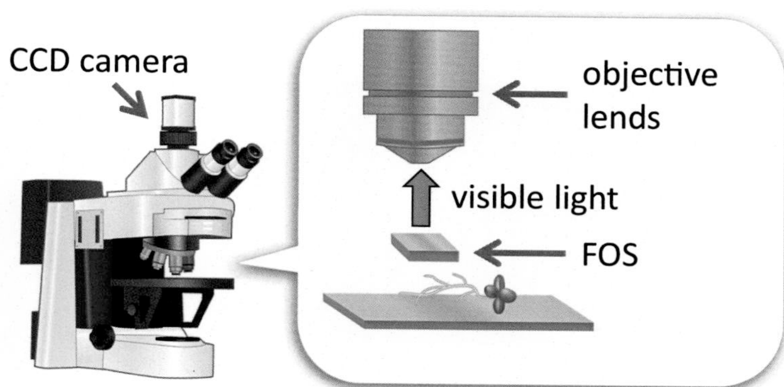

Fig. 19.9 Microscope installed with the RRIS. The radiation from the nuclides in a sample was converted to visible light by FOS and CCD camera captured the light passed through the objective lends

type II system is used to analyze low energy beta emitters such as ^{14}C, ^{35}S, and ^{45}Ca as well as the higher energy beta emitters. However, darkness is still necessary during acquisition by the photon counting camera, which influences plant growth. Therefore, a shorter acquisition time would be better during the daytime.

In addition to the two systems, in order to observe in greater detail, we have developed a RRIS fluorescent microscopy system (Kobayashi et al. 2012) to provide sub-millimeter resolution and to acquire images of radionuclides, as well as fluorescence, chemiluminescence and differential interference contrast from a single sample (Fig. 19.9).

19.3.2 Example: Real-Time Radioisotope Imaging System for Rice in a Paddy Soil

To analyze the Cs movement around the roots (rhizosphere) in a rice plant, we performed a ^{137}Cs tracer experiment with paddy soil from Fukushima and rice plants (Fig. 19.10). We set the rice plant with or without soil in a chamber and collected sequential images every 10 min. The rice plant without soil absorbed ^{137}Cs from the roots and transported it to the leaves within 4 h after ^{137}Cs addition, whereas a signal in the leaves of the rice plant with soil was detected 16 h after adding ^{137}Cs, indicating that Cs fixation in the soil plays an important role to reduce Cs in the plant.

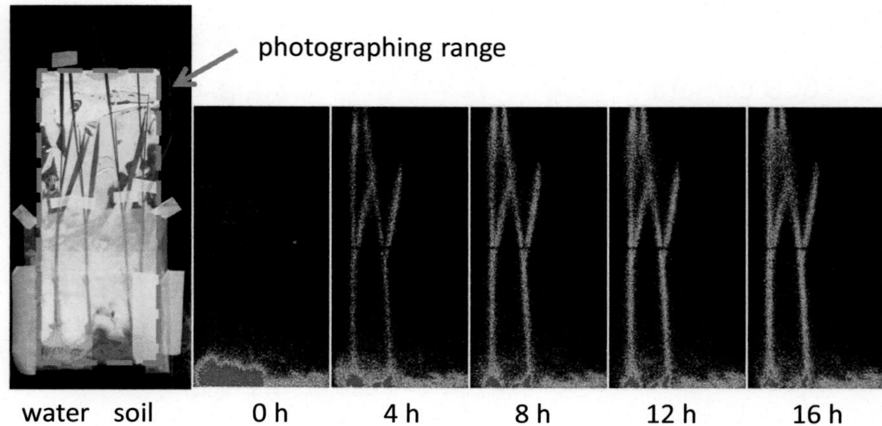

Fig. 19.10 The ^{137}Cs distribution in rice plants grown in a culture solution or paddy soil. The ^{137}Cs was added in a culture solution or paddy soil. The images were captured every 10 min

19.4 Potassium-42

Cesium and K are both monovalent cations in a solution and their movement in animals and plants have been reported to be similar (Zhu and Smolders 2000; Hampton et al. 2005). A tracer experiment was needed to record the ions; however, there was no available isotope of potassium with a long half-life. Potassium-38 has a half-life of 7.6 min, can be made by a cyclotron, and can be applied to plants (Tanoi et al. 2005); however, it is impossible to record it for several hours. Candidates for plant analysis are ^{42}K (half-life is 12.4 h) and ^{43}K (half-life is 22.6 h), which are generally made by a cyclotron (Clark et al. 1972) but not readily available. In order to use a more convenient potassium isotope, a ^{42}Ar–^{42}K generator was used (Wegmann et al. 1981). Since the half-life of ^{42}Ar is about 33 years, the generator can provide ^{42}K (a daughter nuclide of ^{42}Ar) for several decades by milking from the ^{42}Ar–^{42}K generator. The ^{42}Ar–^{42}K generator was used commercially in Japan in the 1980s and now they are used for plant research. We introduce the use of ^{42}K for researching potassium movement in a rice plant.

19.4.1 Methods

^{42}K was obtained from the ^{42}Ar–^{42}K generator using the method of Homareda (Homareda and Matsui 1986) with minor modifications. In this generator, ^{42}Ar, produced by the ^{40}Ar(t, p)^{42}Ar reaction in a cyclotron (Wegmann et al. 1981), decays to continuously produce ^{42}K. A steel cathode is inserted into the generator and (a tension of) approximately 60 V is applied to accumulate the positively charged ^{42}K (Fig. 19.11). The cathode was then washed with a solution to extract the $^{42}K^{+}$. The radioactivity of the extracted solution was measured using a liquid scintillation counter (LSC 6100; ALOKA).

19.4.2 Examples

The movement of ^{42}K in a rice plant was visualized using RRIS as well as in an Arabidopsis (Aramaki et al. 2015). Two-weeks-old seedlings were placed in a RRIS chamber with 3 ml of culture solution containing 6 kBq of ^{42}K. The ^{42}K signal from the sample was recorded every 15 min for 12 h and the signals were calculated by Image J software taking the half-life into account (Fig. 19.12). The signals coming from both the culture solution and the rice plants were observed by the system: this revealed that the rice plant absorbed most of the ^{42}K in the culture solution in 3 h. Using concurrent measurements, the movements of ^{42}K and ^{137}Cs were compared.

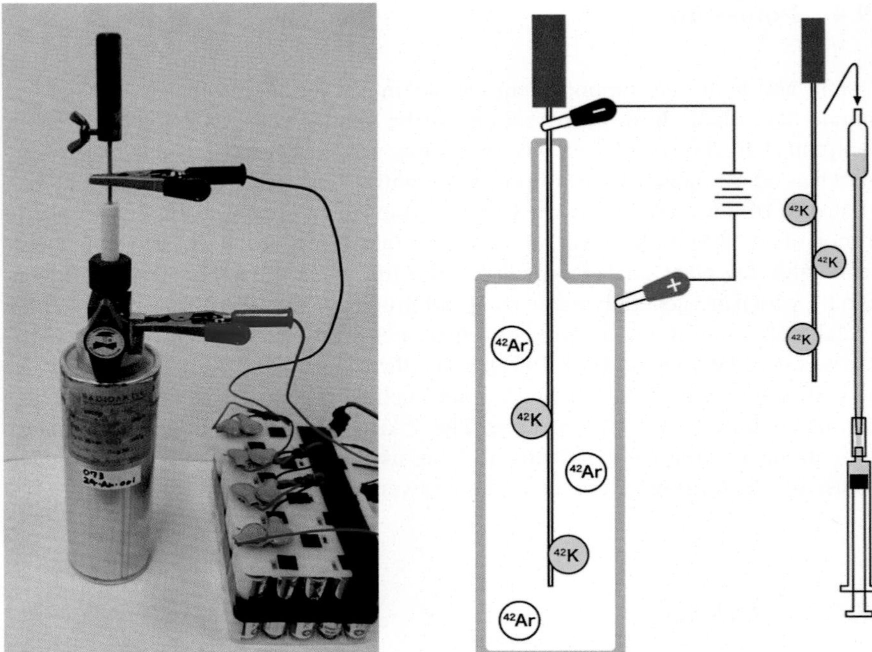

Fig. 19.11 The ^{42}Ar–^{42}K generator. A charge of 60 V was applied to the cathode to attract ionized ^{42}K. A few days after, the extracting solution was filled into the pipette in which the cathode was placed to obtain the ^{42}K

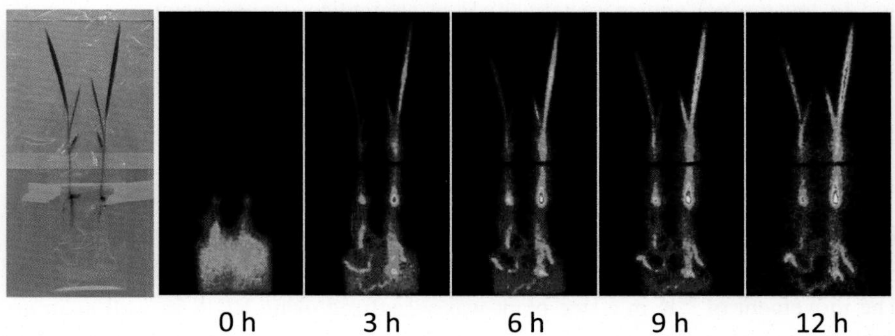

Fig. 19.12 ^{42}K distribution in a rice plant observed by RRIS. The images were captured every 15 min and the shown images were calculated considering the half life of ^{42}K

19.5 Summary

We introduced some new techniques for visualizing radionuclides including radiocesium. There are several choices of imaging techniques. To observe the radiocesium distribution in a thin sample, RLG with an IP would be the most convenient technique. When it is necessary to observe the radiocesium distribution in detail, frozen section imaging with RLG or MAR are useful for the distribution can be observed at a resolution of tens to hundreds of micrometers. As a live imaging system, RRIS can contribute to understanding the movement of radiocesium in plants. With these imaging techniques, as well as with nuclides such as ^{137}Cs, ^{42}K, and ^{22}Na, we can access important information to solve the problem of radiocesium contamination in crops, plants, and the environment.

Acknowledgement The authors thank Xerri, Sophie-Asako for critical reading of the manuscript

References

Aramaki T, Sugita R, Hirose A et al (2015) Application of ^{42}K to Arabidopsis tissues using Real-Time Radioisotope Imaging System (RRIS). Radioisotopes 64:169–176. doi:10.3769/radioiso topes.64.169

Clark JC, Thakur ML, Watson IA (1972) The production of potassium-43 for medical use. Int J Appl Radiat Isot 23:329–330. doi:10.1016/0020-708X(72)90011-7

Fujimaki S, Suzui N, Ishioka NS et al (2010) Tracing cadmium from culture to spikelet: noninvasive imaging and quantitative characterization of absorption, transport, and accumulation of cadmium in an intact rice plant. Plant Physiol 152:1796–1806. doi:10.1104/pp. 109. 151035

Furuta E (2013) Semi-quantitative analysis of leaf surface contamination by radioactivity from the Fukushima Daiichi nuclear power plant accident using HPGe and imaging plate. J Radioanal Nucl Chem 297:337–342. doi:10.1007/s10967-012-2343-4

Hampton C, Broadley M, White P (2005) Short review: the mechanisms of radiocaesium uptake by *Arabidopsis* roots. Nukleonika 50(suppl1):3–8

Hirose A, Kobayashi NI, Tanoi K, Nakanishi TM (2014) A microautoradiographic method for fresh-frozen sections to reveal the distribution of radionuclides at the cellular level in plants. Plant Cell Physiol 55:1194–1202. doi:10.1093/pcp/pcu056

Homareda H, Matsui H (1986) Biochemical utilization of ^{42}Ar-^{42} K generator. Radioisotopes 35:543–546. doi:10.3769/radioisotopes.35.10_543

Isobe T, Mori Y, Takada K et al (2013) Evaluation of vegetables in Tsukuba for contamination with radioactive materials from the accident at Fukushima Daiichi nuclear power plant. Health Phys 105:311–317. doi:10.1097/HP.0b013e3182895759

Itoh S, Eguchi T, Kato N, Takahashi S (2014) Radioactive particles in soil, plant, and dust samples after the Fukushima nuclear accident. Soil Sci Plant Nutr 60:540–550. doi:10.1080/00380768. 2014.907735

Jahnke S, Menzel MI, van Dusschoten D et al (2009) Combined MRI-PET dissects dynamic changes in plant structures and functions. Plant J 59:634–644. doi:10.1111/j.1365-313X.2009.03888.x

Kanno S, Rai H, Ohya T et al (2007) Real-time imaging of radioisotope labeled compounds in a living plant. J Radioanal Nucl Chem 272:565–570. doi:10.1007/s10967-007-0625-z

Kobayashi NI, Tanoi K, Kanno S, Nakanishi TM (2012) Analysis of the iron movement in the root tip part using real-time imaging system. Radioisotopes 61:121–128. doi:10.3769/radioisotopes.61.121

Koizumi A, Niisoe T, Harada KH et al (2013) ^{137}Cs trapped by biomass within 20 km of the Fukushima Daiichi nuclear power plant. Environ Sci Technol 47:9612–9618. doi:10.1021/es401422g

Mimura T, Mimura M, Kobayashi D et al (2014) Radioactive pollution and accumulation of radionuclides in wild plants in Fukushima. J Plant Res 127:5–10. doi:10.1007/s10265-013-0599-6

Mori S, Hirato A, Tanoi K et al (2012) Radioactive cesium flow in *Rhus vernicifera*. Soil Sci Plant Nutr 58:611–617. doi:10.1080/00380768.2012.727188

Mukai H, Hatta T, Kitazawa H et al (2014) Speciation of radioactive soil particles in the Fukushima contaminated area by IP autoradiography and microanalyses. Environ Sci Technol 48:13053–13059. doi:10.1021/es502849e

Nakajima H, Fujiwara M, Tanihata I et al (2014) Imaging plant leaves to determine changes in radioactive contamination status in Fukushima, Japan. Health Phys 106:565–570

Nakanishi T, Yamawaki M, Kannno S et al (2009) Real-time imaging of ion uptake from root to above-ground part of the plant using conventional beta-ray emitters. J Radioanal Nucl Chem 282:265–269. doi:10.1007/s10967-009-0343-9

Nakanishi TM, Kobayashi NI, Tanoi K (2012) Radioactive cesium deposition on rice, wheat, peach tree and soil after nuclear accident in Fukushima. J Radioanal Nucl Chem. doi:10.1007/s10967-012-2154-7

Nakanishi H, Tanaka H, Takeda K et al (2014) Radioactive cesium distribution in bamboo [*Phyllostachys reticulata*(Rupr) K. Koch] shoots after the TEPCO Fukushima Daiichi nuclear power plant disaster. Soil Sci Plant Nutr 60:801–808. doi:10.1080/00380768.2014.939936

Niimura N, Kikuchi K, Tuyen ND et al (2014) Physical properties, structure, and shape of radioactive Cs from the Fukushima Daiichi Nuclear Power Plant accident derived from soil, bamboo and shiitake mushroom measurements. J Environ Radioact. doi:10.1016/j.jenvrad.2013.12.020

Sakamoto F, Ohnuki T, Kozai N et al (2013) Determination of local-area distribution and relocation of radioactive cesium in trees from Fukushima Daiichi nuclear power plant by autoradiography analysis. Trans At Energy Soc Jpn 12:257–266

Shiba K, Kitamura Y, Kozaka T et al (2013) Decontamination of radioactivity from contaminated vegetables derived from the Fukushima nuclear accident. Radiat Meas 55:26–29. doi:10.1016/j.radmeas.2013.01.010

Sugita R, Kobayashi NI, Hirose A et al (2014) Evaluation of in vivo detection properties of ^{22}Na, ^{65}Zn, ^{86}Rb, ^{109}Cd and ^{137}Cs in plant tissues using real-time radioisotope imaging system. Phys Med Biol 59:837–851. doi:10.1088/0031-9155/59/4/837

Suzuki T, Mori C, Yanagida K et al (1997) Characteristics and correction of the fading of imaging plate. J Nucl Sci Technol 34:461–465. doi:10.1080/18811248.1997.9733692

Takata D, Yasunaga E, Tanoi K et al (2012) Radioactivity distribution of the fruit trees ascribable to radioactive fall out (III): a study on peach and grape cultivated in south Fukushima. Radioisotopes 61:601–606

Tanaka K, Iwatani H, Sakaguchi A et al (2013) Local distribution of radioactivity in tree leaves contaminated by fallout of the radionuclides emitted from the Fukushima Daiichi Nuclear Power Plant. J Radioanal Nucl Chem 295:2007–2014. doi:10.1007/s10967-012-2192-1

Tanoi K (2013) Behavior of radiocesium adsorbed by the leaves and stems of wheat plant during the first year after the Fukushima Daiichi nuclear power plant accident. In: Nakanishi TM,

Tanoi K (eds) Agricultural implications of the Fukushima nuclear accident. Springer Japan, Tokyo, pp 11–18

Tanoi K, Hojo J, Suzuki K et al (2005) Analysis of potassium uptake by rice roots treated with aluminum using a positron emitting nuclide, K-38. Soil Sci Plant Nutr 51:715–717

Tanoi K, Hashimoto K, Sakurai K et al (2011) An imaging of radioactivity and determination of an imaging of radioactivity and determination of Cs-134 and Cs-137 in wheat tissue grown in Fukushima. Radioisotopes 60:317–322. doi:10.3769/radioisotopes.60.299

Wegmann H, Huenges E, Muthig H, Morinaga H (1981) Acceleration of tritons with a compact cyclotron. Nucl Instrum Methods 179:217–222. doi:10.1016/0029-554X(81)90042-2

Yamaguchi T, Sawano K, Furuhama K et al (2013) An autoradiogram of skeletal muscle from a pig raised on a farm within 20 km of the Fukushima Daiichi nuclear power plant. J Vet Med Sci 75:93–94

Yamawaki M, Hirose A, Kanno S et al (2010) Evaluation of ^{109}Cd detection performance of a real-time RI imaging system for plant research. Radioisotopes 59:155–162

Zeissler CJ, Forsley LPG, Lindstrom RM et al (2013) Radio-microanalytical particle measurements method and application to Fukushima aerosols collected in Japan. J Radioanal Nucl Chem 296:1079–1084. doi:10.1007/s10967-012-2135-x

Zhu YG, Smolders E (2000) Plant uptake of radiocaesium: a review of mechanisms, regulation and application. J Exp Bot 51:1635–1645

Printed in the United States
By Bookmasters